Forensic
Mass
Spectrometry

Editor
Jehuda Yinon, Ph.D.
Department of Isotope Research
Weizmann Institute of Science
Rehovot, Israel

CRC Press, Inc.
Boca Raton, Florida

Library of Congress Cataloging-in-Publication Data

Forensic mass spectrometry.

1. Chemistry, Forensic--Technique. 2. Mass spectrometry--Forensic applications. I. Yinon, Jehuda.
[DNLM: 1. Forensic Medicine--methods. 2. Spectrum Analysis, Mass--methods. W 750 F7125]
HV8073.5.F67 1987 363.2′562 87-10362
ISBN 0-8493-5366-1

To the memory of my parents Moshe and Bluma Fischler

"Justice is truth in action"

Benjamin Disraeli, 1804—1881
(Speech, House of Commons, February 11, 1851)

INTRODUCTION

The word "forensic" originates from the Roman period, when the market place—forum—served also as court of justice.

Forensic science applies the principles of natural science to the purposes of justice. The task of the forensic scientist (criminalist, chemist, or toxicologist) can be summed up as identification, comparison, and individualization in the following:

1. Analysis of evidence, which includes the identification of drugs, explosives, hair, paint, glass, fibers, flammable materials, and body fluids. Also, the detection and identification of drugs and poisons in body fluids.
2. Comparison of evidence materials in order to link evidence from the scene of a crime to a suspect. For example, the comparison of paint, hair, fiber, glass, soil, bullets,s fingerprints, or bloodstains.
3. Individualization or authentication of ancient objects, pieces of art, precious stones, etc., in order to prevent and/or detect forgeries. Also, the detection of forgery of documents and currency.

The continued improvements in anlaytical instrumentation during the last 20 years has had a major influence on the forensic laboratory. The analytical methods in use include thin-layer chromatography (TLC), gas chromatography (GC), high performance liquid chromatography (HPLC), ultraviolet (UV) and infrared (IR) spectroscopy, and mass spectrometry.

Mass spectrometry, in particular as a combination with gas chromatography (GC/MS) and with liquid chromatography (LC/MS), had a major impact on the forensic laboratory. The use of the mass spectrometer, interfaced with an on-line mini- or microcomputer in an interactive mode, has added a new dimension to forensic science analytical instrumentation; the kinds of substances that can be analyzed have increased, sensitivity of detection has been enhanced, the ability to isolate and identify minor constituents in complex mixtures has been improved, and the time required for an analysis has diminished. GC/MS has been for some time a widely used method for the analysis of volatile substances. The introduction of coated capillary columns, and the use of alternate ionization techniques, like electron impact (EI) and chemical ionization (CI), have made it the method of choice in the forensic laboratory. With the development of new interfacing techniques, LC/MS has become complimentary to GC/MS, especially for less volatile mixtures. Tandem mass spectrometry (MS/MS) can provide an extra dimension of structural information for pure compounds and can also serve as a fast separation and identification method for mixtures.

The mass spectrometer can also be used to provide the forensic scientist with information about solids: pyrolysis-mass spectrometry, field desorption (FD) and fast atom bombardment (FAB) ionization for organic solids, and thermal ionization and spark source mass spectrometry for inorganic solids.

It is the purpose of this book to bring together the available information on the use of mass spectrometry in forensic science. The subjects of the various chapters were chosen not only with respect to their importance in forensic science, but also according to the contribution of mass spectrometry — in the present and future — as a major analytical technique in that particular field.

Mass spectrometry, and in particular CI, has played a major role in forensic identification of commonly abused drugs. The anlysis of drugs and toxic substances in body fluids, or toxicology, is today largely based on GC/MS. The use of mass spectrometry in drug control in sports received a large boost during the 1984 Olympic games in Los Angeles, when a series of GC/MS instruments were used for drug testing of urine samples of athletes. Equine drug testing in horse races has been done for some time by GC/MS and LC/MS. No doubt that these methods will become routine for drug testing at major sporting events. Mass spectrometry has become in many forensic laboratories part of the routine analytical procedure for the identification of postexplosion residues. It has also been incorporated in several "sniffers" for the detection of hidden explosives. GC/MS has proved to be useful in suspected arson cases, to provide evidence of a volatile accelerant used to start a fire. Pyrolysis-mass spectrometry has been used sos far by only a few forensic laboratories. It is, however, a promising technique for the characterization of synthetic polymers. Mass spectrometry has also been used for the analysis and characterization of a variety of substances of forensic interest. Several examples are presented in the chapter on miscellaneous applications.

The mass spectrometer has become a well-established technique in the forensic laboratory. The expansion of its range of applications depends now on the ingenuity of the forensic scientist.

THE EDITOR

Jehuda Yinon, Ph.D. is a Senior Research Fellow and Head of the Applied Mass Spectrometry Group at the Weizmann Institute of Science, Rehovot, Israel. His main research interests are applications of novel mass spectrometry techniques in forensic science, the chemistry of energetic materials and the study of metabolism of explosives. He has a B.Sc. and M.Sc. from the Technion, Israel Institute of Technology, Haifa, Israel and a Ph.D. from the Weizmann Institute of Science. He was a Research Associate (1971 to 1973) and a Senior Research Associate (1976 to 1977) at the Jet Propulsion Laboratory, Pasadena, California and a Visiting Scientist (1981 to 1982) at the National Institute of Environmental Health Sciences in North Carolina.

Dr. Yinon is an associate editor of *Critical Reviews in Forensic Sciences* and a member of the editorial board of the *Journal of Energetic Materials*. He is currently the President of the Israel Society for Mass Spectrometry.

CONTRIBUTORS

Jack D. Henion, Ph.D.
Associate Professor
Equine Drug Testing and Toxicology
NYS College of Veterinary Medicine
Cornell University
Ithaca, New York

Michael Klein, Ph.D.
Senior Chemist
Drug Control Section
Drug Enforcement Administration
Washington, D.C.

Ray H. Liu, Ph.D.
Associate Professor
Department of Criminal Justice
School of Social and Behavioral Science
University of Alabama
Birmingham, Alabama

George Maylin, Ph.D.
Associate Professor
Drug Testing and Toxicology
NYS Equine College of Veterinary
  Medicine
Cornell University
Ithaca, New York

Dominique Silvestre
Research Associate
Drug Testing and Toxicology
NYS College of Veterinary Medicine
Cornell University
Ithaca, New York

R. Martin Smith, Ph.D.
Quality Control Chemist
Wisconsin State Crime Laboratory
Madison, Wisconsin

Brian B. Wheals
Senior Principal Scientific Officer
Special Services Section
Metropolitan Forensic Science Laboratory
London, England

Michael John Whitehouse, Ph.D.
Principal Scientific Officer
Special Services Section
Metropolitan Police Forensic Science
  Laboratory
London, England

Jehuda Yinon, Ph.D.
Senior Research Fellow
Department of Isotope Research
Weizmann Institute of Science
Rehovot, Israel

TABLE OF CONTENTS

Chapter 1

MASS SPECTROMETRY OF COMMONLY ABUSED DRUGS

Ray [J.] H. Liu

TABLE OF CONTENTS

I. INTRODUCTION

Mass spectrometry (MS), in combination with an appropriate separation method, is undoubtedly the single most powerful tool for the analysis of organic compounds. As a result of its recent availability in most forensic laboratories, MS-based techniques have become the methods of choice in most drug-related analyses. General references in this aspect[1-5] and data compilation,[6-8] specific for forensic drug analysis needs, are now available in the literature.

A. Scope and Organization

Comprehensive MS literature with emphasis on basic understanding, application, and data compilation are readily available. Interested readers should consult these sources for general and detailed information. This chapter will address only the analytical aspect of MS as applied to the analysis of commonly abused drugs. The analysis of drugs of abuse and their metabolites in body fluids constitutes a separate entity and is the topic of Chapter 2. Only the analysis of these drugs in "dosage" or raw material forms will be treated in this chapter. Occasionally, references mainly addressing the analysis of drugs (and their metabolites) in body fluids may be cited if they also contain pertinent information related to the topics of this chapter.

The term "drugs of abuse" cannot be definitely defined.[9] For the purpose of this chapter, drugs that are not categorized as one of the five schedules[10] of the Federal Controlled Substances Act will not be included. Drugs commonly manufactured in clandestine laboratories[11,12] are particularly emphasized.

Literature falling within the hereby described scope are reviewed and grouped, according to their chemical categories, in Section III. Within each drug category, electron impact (EI) studies are first summarized, followed by other ionization, mainly chemical ionization (CI), methods, and sample introduction mechanisms. Several interesting approaches as applied to drug analysis are summarized in Section IV, which also includes the applications of the still-developing tandem MS (MSMS) to the analysis of drugs of forensic interest.

B. Controlled Substances

The Federal Controlled Substances Act, enacted in 1970, established a comprehensive pattern of control over the manufacture and distribution of drugs. The Act set up five schedules upon which different criminal penalties are placed for unlawful trafficking. Possession for one's own use of any controlled substance in any schedule is always a misdemeanor on the first offense, punishable by 1 year in jail and up to a $5000 fine.

The scheduling of a drug is based on (1) its actual or relative potential for abuse, (2) scientific evidence of its pharmacological effect, if known, (3) the state of current scientific knowledge regarding the drug or other substance, (4) its history and current pattern of abuse, (5) the scope, duration, and significance of abuse, (6) what, if any, risk there is to the public health, (7) its psychic or physiological dependence liability, and (8) whether the substance is an immediate precursor of a substance already under control.[13] Table 1 summarizes the scheduled drugs listed in the Code of Federal Regulation.[10]

II. SELECTION OF IONIZATION AND SAMPLE INTRODUCTION METHODS

Despite the availability of several "soft ionization" methods, CI is the only one which has been used to any significant extent in the analysis of drug samples.[14] Since

Table 1
SCHEDULE OF CONTROLLED SUBSTANCES

Schedule I

Opiate and opium derivative

Acetorphine	Acetyldihydrocodeine	Acetylmethadol
Allylprodine	Alphacetylmethadol	Alphameprodine
Alphamethadol	Alphamethylfentanyl	Benzethidine
Benzylmorphine	Betacetylmethadol	Betameprodine
Betamethadol	Betaprodine	Clonitazene
Codeine methylbromide	Codeine-N-oxide	Cyprenorphine
Desomorphine	Dextromoramide	Diampromide
Diethylthiambutene	Difenoxin	Dihydromorphine
Dimenoxadol	Dimepheptanol	Dimethylthiambutene
Dioxaphetyl butyrate	Dipipanone	Drotebanol
Ethylmethylthiambutene	Etonitazene	Etorphine
Etoxeridine	Furethidine	Heroin
Hydromorphinol	Hydroxypethidine	Ketobemidone
Levomoramide	Levophenacylmorphan	Methyldesophine
Methyldihydromorphine	Morpheridine	Morphine methylbromide
Morphine methylsulfonate	Morphine-N-oxide	Myrophine
Nicocodeine	Nicomorphine	Noracymethadol
Norlevorphanol	Normethadone	Normorphine
Norpipanone	Phenadoxone	Phenampromide
Phenomorphan	Phenoperidine	Pholcodeine
Piritramide	Proheptazine	Properidine
Propiram	Racemoramide	Sufentanil
Thebacon	Tilidine	Trimeperidine

Hallucinogenic substance

4-Bromo-2,5-dimethoxyam-phetamine	2,5-Dimethoxyamphetamine	4-Methoxyamphetamine
3,4-Methylenedioxy ampheta-mine	5-Methoxy-3,4-methylene-dioxy-amphetamine	5-Methyl-2,5-dimethoxyam-phetamine
Diethyltryptamine	3,4,5-trimethoxy ampheta-mine	Bufotenine
Ibogaine	Lysergic acid diethylamide	Dimethyltryptamine
Mescaline	N-Ethyl-3-piperidyl benzilate	Marijuana
Peyote	Tetrahydrocannadinols	Parahexyl
Psilocybin	Thiophene analog of phency-clidine	N-Methyl-3-piperidyl benzi-late
Psilocyn		
Pyrrolidine analog of phency-clidine		Ethylamine analog of phency-clidine

Depressant

Mecloqualone

Stimulant

Fenethylline	N-Ethylamphetamine

Table 1 (continued)

SCHEDULE OF CONTROLLED SUBSTANCES

Schedule II

Opium, opiate, and other substance of vegetable origin

Alphaprodine	Anileridine	Bezitramide
Cocaine and equivalents	Coca leaf	Codeine
4-Cyano-2-dimethylamino-4,4-diphenyl butane	4-Cyano-1-methyl-4-phenyl-piperidine	Dextropropoxyphene (bulk)
Dihydrocodeine	Diphenoxylate	Ecgonine and equivalents
Ethylmorphine	Ethyl-4-phenylpiperidine-4-carboxylate	Etorphine hydrochloride
Fentanyl	Isomethadone	Hydrocodone
Hydromorphone	Methadone	Levophanol
Metazocine	Metopon	1-Methyl-4-phenylpiperidine-4-carboxylic acid
2-Methyl-3-morpholino-1,1-diphenylpropane-carboxylic acid	Morphine	Opium (extract, granulated, powdered, raw, tincture)
Oxycodone	Opium poppy	Poppy straw (raw, concentrate)
Pethidine	Oxymorphone	Racemorphan
Piminodine	Phenazocine	
Thebaine	Racemethorphan	

Stimulant and precursor

Amphetamine (its salts, optical isomers and salts of optical isomers)	Methamphetamine (its salt, isomers and salts of isomers)	Methylphenidate
	Phenylacetone	Phenmetrazine and its salts

Depressant and precursor

Amobarbital	Methaqualone	Pentobarbital
Phencyclidine	Phenylcyclohexylamine	Piperidinocyclohexanecarbonitrile
Secobarbital		

Schedule III

Narcotic drug

Any material containing less than a specified quantity of the following drugs: codeine, dihydrocodeinone, dihydrocodeine, ethylmorphine, morphine, or opium

Stimulant

Benzphetamine	Chlorphentermine	Clotermine
Phendimetrazine	Schedule II stimulants in dosage unit form	

Depressant

Chlorhexadol	Glutethimide	Lysergic acid
Lysergic acid amide	Methyprylon	Sulfondiethylmethane
Sulfonmethane		

Substances containing amobarbital, secobarbital, or pentobarbital

Substances containing a derivative of barbituric acid or any salt thereof

Table 1 (continued)
SCHEDULE OF CONTROLLED SUBSTANCES

Other

Nalorphine

Schedule IV

Narcotic drug

Dextropropoxyphene
Any substance containing specified quantities of difenoxin and atropine sulfate

Stimulant

Diethylpropion	Mazindol	Pemoline
Phentermine	Pipradrol	(−)-1-Dimethylamino-1,2-di-phenylethane

Depressant

Alprazolam	Barbital	Chloral betaine
Chloral hydrate	Chlordiazepoxide	Clonazepam
Chlorazepate	Diazepam	Ethchlorvynol
Ethinamide	Flurazepam	Halazepam
Lorazepam	Mebutamate	Meprobamate
Methohexital	Methylphenobarbital	Oxazepam
Paraldehyde	Petrichloral	Phenobarbital
Prazepam	Temazepam	Triazolam

Other

Fenfluramine	Pentazocine

Schedule V

Any substance containing specified quantities of nonnarcotic active medicinal ingredient and the following narcotic drug: codeine, dihydrocodeine, ethylmorphine, diphenoxylate, opium, or defenoxin

volatility is rarely a problem for drug samples, solid-state ionization techniques, such as laser desorption, do not normally offer significant advantage for this application. It is interesting, though, to note the exploration of using secondary ion MS for the analysis of drugs in salt forms,[15] and the use of a ^{63}Ni foil at atmospheric pressure for sample ionization.[16] Both positive and negative ion mass spectra are obtained for cocaine and barbiturates in the latter study.

Spectra obtained by EI, positive (PCI) and negative CI (NCI) and charge exchange (CE) ionization methods are compared in several studies.[17-22] Figure 1 summarizes these comparisons. Since these spectra are obtained under different conditions in different laboratories, quantitative comparison may not be valid. However, their comparative values in revealing molecular weight or structural information are apparent. It should be noted that CI often generates a certain degree of fragmentation and may also supplement structural information[23] provided by EIMS. The highest mass ion with significant intensity resulting from NCI is often, but not always, derived from the loss of one (sometimes two or more) hydrogen ion from the molecular anion.[21] It appears that spectra produced by this ionization technique are less predictable (or less under-

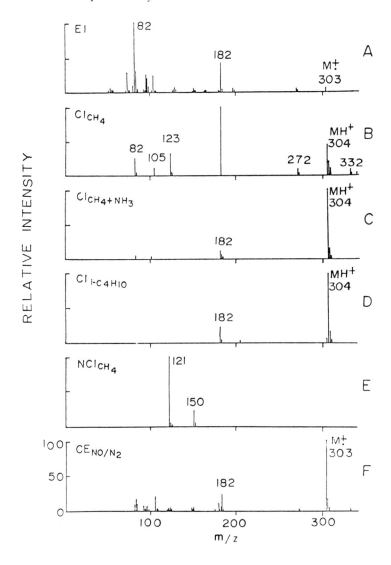

FIGURE 1.   Mass spectra of cocaine obtained by different ionization methods. A: EI, B: CI (methane), C: CI (methane + ammonia), D: CI (isobutane), E: negative CI (methane), F: CE (10% nitric oxide in nitrogen). (D is constructed based on a table in Milne, G. W. A., Fales, H. M., and Axenrod, T., *Anal. Chem.,* 43, 1815, 1971. A, B, and C are from Foltz, R. L., Fentiman, A. F., Jr., and Foltz, R. B., *GC/MS Assays for Abused Drugs in Body Fluids,* U.S. Government Printing Office, Washington, D.C., 1980; E is from Foltz, R. L., *Quantitative Mass Spectrometry in Life Sciences II,* de Leenheer, A. P., Foncucci, F. F., and van Peteghem, C., Eds., Elsevier, Amsterdam, 1978; F is from Jardine, I. and Fenselau, C., *Anal. Chem.,* 47, 730, 1975. With permission.)

stood) compared to its positive counterpart. However, enhanced sensitivities are often achievable when compounds with high electron affinities are analyzed.[24] It is interesting to note that ammonium carbonate and sodium bicarbonate have been used for the *in situ* generation of reagent gases in PCI[25] and NCI[26,27] methods which include the analysis of drugs.

Although MSMS has been successfully used in the analysis of complex mixtures of forensic interest,[28-34] a chromatograph is usually used as the sample inlet device. The applications of gas chromatography/mass spectrometry (GCMS) to the analysis of drug samples in forensic laboratories has been well demonstrated.[35-37] The advantages of using CI techniques have also been emphasized.[38] In complement to gas chromatographs, liquid chromatographs are often used to introduce drug samples not suitable for gas chromatographic conditions.[39-41] Pyrolysis GC[42,43] and thin-layer chromatographic (TLC) plates[44,45] have also been used as sample introduction devices, but these applications are rare and probably have limited potential.

III. ANALYSIS OF COMMONLY ABUSED DRUGS

A. Morphine and Related Alkaloids

Due to their pharmaceutical significance, morphine alkaloids have been well studied,[46-49] with emphasis on the fragmentation mechanisms under EI conditions. Fragmentation pathways are established mainly based on the observed mass shifts, with respect to morphine, of morphine-related compounds. It is concluded that major fragments derive from the cleavage of the two bonds in β-position to the nitrogen atom. Major fragmentation mechanisms are summarized in Figure 2.[49,50] It should be noted that compounds without the ether ring undergo much more rapid fragmentation.[46] It has also been reported that the relative abundance of the molecular ion and the m/z 59 ion in morphine (or its analogs in other compounds) depend on the stereochemistry (*cis-* or *trans-*) of the B and C rings in the compound under examination.[48] CIMS studies[23,51-56] of morphine alkaloids have been undertaken mainly for analytical purposes, with emphasis being placed on the simplicity of the CI spectra. This advantage allows the identification of heroin in the presence of its diluents[54] and the analysis of raw opium.[55] It has also been used to study an extended list of morphine-related compounds.[56] Since spectra obtained from conventional CIMS often generate base peaks other than the $(M + H)^+$ ion, CIMS in the form of CE, using 10% NO in N_2, is investigated.[18,19] Relative intensities of the molecular ions (or "pseudo-molecular ion") observed under EI, isobutane CI, and 10% NO (in N_2) CE ionization methods are summarized in Table 2. These data show that, under CE conditions, the molecular ions are the base peaks in all 12 compounds studied. This is only true for five compounds under EI and isobutane CI conditions.

It should be noted that certain unsaturated compounds undergo a reduction reaction[57-59] under CI conditions, thus producing protonated molecular ions of the reaction products. This phenomenon is observed for codeinone,[57] and should be recognized when qualitative and quantitative analysis of unsaturated drugs are addressed.

In order to probe the number and the nature of active hydrogens attached to heteroatoms in the compounds under examination, various deuterated reagent gases have been used in CIMS.[59,60] This approach reveals more structural information and increases the analytical value of CIMS. Based on the mass shifts of the "protonated" ion observed when switching from ammonia to d_3-ammonia, the number of active hydrogens in morphine and several other drugs are well demonstrated.[60] The combined information provided by (1) mass shifts, (2) the tendencies of the protonated and adduct ions to lose molecules of water, and (3) the adduct ion to protonated ion intensity ratios, further reveals the nature of the active hydrogens involved.[59] To demonstrate the utilization of this information, the CI spectra of ten morphine alkaloids, 1 (morphine), 2 (norcodeine), 3 (hydromorphone), 4 (codeine), 5 (hydrocodone), 6 (thebaine), 7 (isothebaine), 8 (normorphine), 9 (codeinone), and 10 (oxycodone), are summarized in Table 3[59] and illustrated as follows.

The total number of active hydrogens in a compound is derived from the mass shifts of the "protonated" ion observed when switching from ammonia to d_3-ammonia:

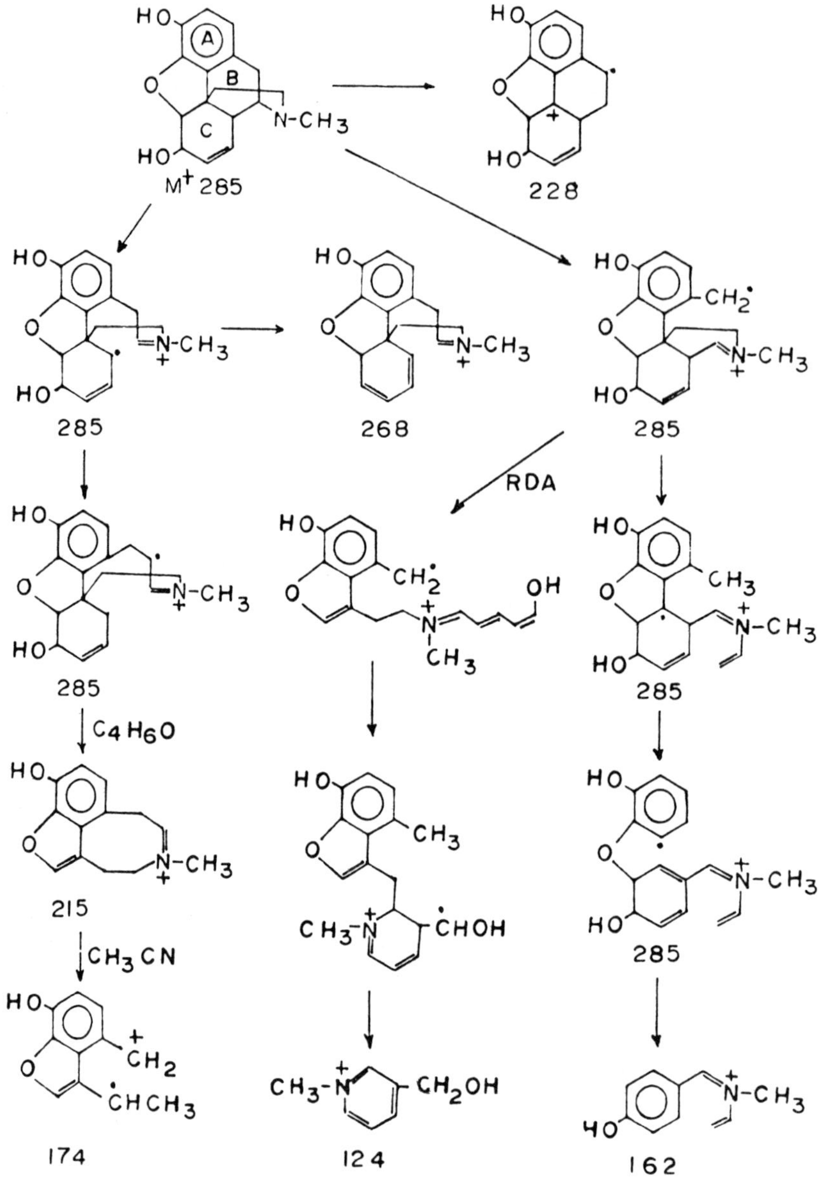

FIGURE 2.   Major EI fragmentation patterns of morphine alkaloids. (From Sastry, S. D., in *Biochemical Applications of Mass Spectrometry*, Waller, G. R., Ed., Wiley-Interscience, New York, 1972, chap. 24. With permission.)

$$n = M_{ND_3} - M_{NH_3} - 1$$

where M_{ND_3} and M_{NH_3} are the mass to charge ratios of the "protonated" ions observed with ND_3 and NH_3 as the reagent gases. Thus, compounds 1 and 2 can be differentiated from compound 3 based on this mass shift as a result of the difference in the number of active hydrogens in these compounds. Under the methane CI conditions used throughout this series of alkaloids, an allyl hydroxyl group generates a $(M + H - 18)^+$ ion intensity three times that of the $(M + H)^+$ ion, while a phenolic OH group generates a $(M + H^+)$ ion intensity 30 times that of the $(M + H - 18)^+$ ion. These $(M + H)^+:(M +$

Table 2
RELATIVE INTENSITIES OF MOLECULAR IONS OF
MORPHINE ALKALOIDS OBSERVED UNDER EI,
ISOBUTANE CI, AND 10% NO/N$_2$ CE MASS
SPECTROMETRY[a]

| | | Relative molecular ion intensity | | |
Compound	Mol wt	EI	CI (M+1)[+]	CE
Heroin	369	25	15	100
O^3-Monoacetylmorphine	327	96	18	100
O^6-Monoacetylmorphine	327	98	20	100
Morphine	285	100	15	100
Nalorphine	311	100	20	100
Codeine	299	100	13	100
Hydrocodone	299	100	100	100
Levorphanol	257	64	100	100
Levallorphan	283	100	100	100
Meperidine	247	22	100	100
Cocaine	303	26	100	100
Atropine	289	15	16	100

[a]   Abstracted from Reference 19.

H − 18)[+] ratios suggest that the hydrogen in compound 3 is phenolic. Furthermore, as a result of differences in proton affinity, compound 3 differs from 1 and 2 by showing a more intense adduct ion intensity with either methane or ammonia CI. Compound 1 differs from compound 2 by showing a substantial intensity (4.3%) of the (M + C$_2$H$_5$ − 18)[+] ion. The two nonalcoholic hydrogens in compounds 1 and 2 cause a small (8.8% vs. 6.1%), but significant difference in their adduct ion to protonated ion intensity ratios. This difference (81% vs. 77%) is again observed when the more basic ammonia is used as the reagent gas. This subtle difference shown by the two different nonalcoholic hydrogens in compounds 1 and 2 undoubtedly will not alert casual observers during routine analysis. On the other hand, when structural elucidations are emphasized and studied under precisely controlled conditions, this difference is recognizable.

Anions in NCI are generally formed through electron capture and reactant-ion CI mechanisms. This technique has been applied to the analysis of morphine and several related compounds.[20,26] In the study of morphine using CO$_2$ as the reagent gas, the most intense peak represents the loss of one hydrogen from the molecular anion. When methane is used, the most intense peak is (M − 2)[-]. Under the methane conditions, the most intense peaks for codeine and dihydrocodeinone are (M − 1)[-] and M[-].

GCMS has been applied to the analysis of morphine, codeine, thebaine, and papaverine in opium samples.[61] It is also useful in the identification of impurities and by-products in illicit heroin preparations.[62,63] Using heptafluorobutyric anhydride as a derivatizing reagent, GCMS has been used to identify the preparation of O^3-monoacetylmorphine in an illicit heroin preparation. Columns packed with OV-17[36,62,63] or OV-101[61] are used in these applications.

B. Amphetamine and Related Amine Drugs

During the 45-month period ending in September 1981, the Drug Enforcement Administration of the U.S. Department of Justice seized a total of 751 clandestine drug laboratories in the U.S. More than half of these clandestine activities involved the manufacture of amine, mainly methamphetamine, drugs.[11,12] Understandably, mass

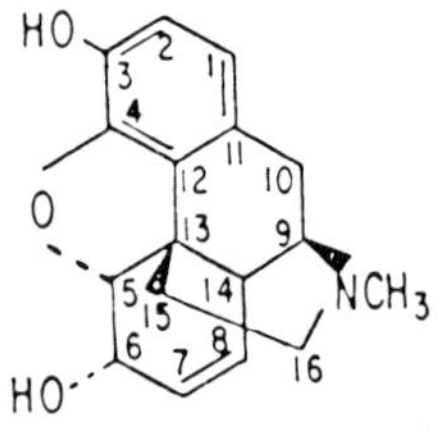

1

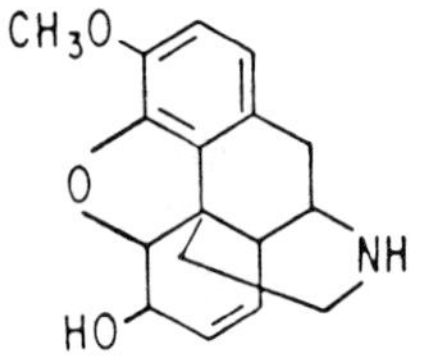

2

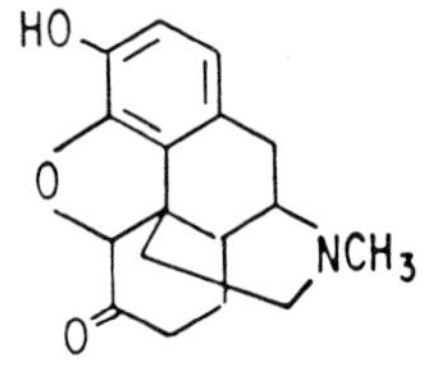

3

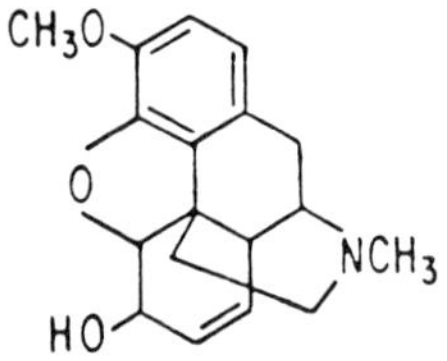

4

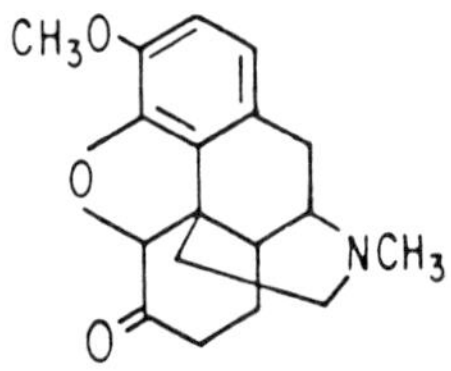

5

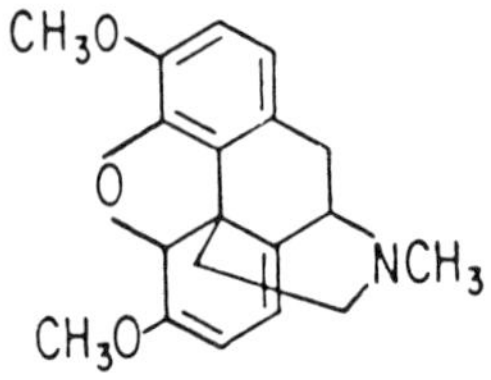

6

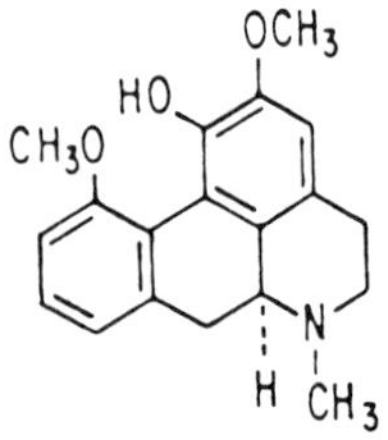

7

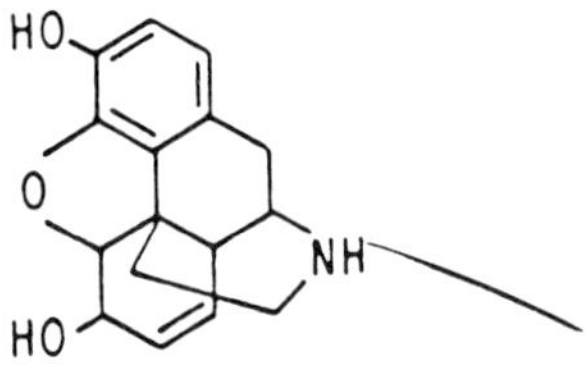

8

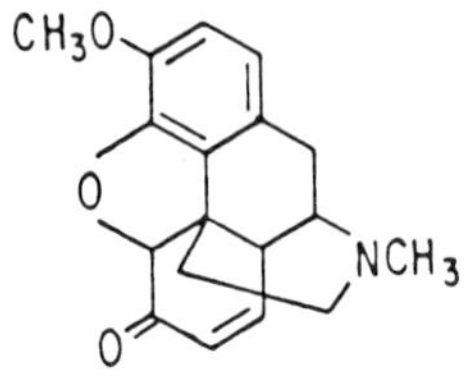

9

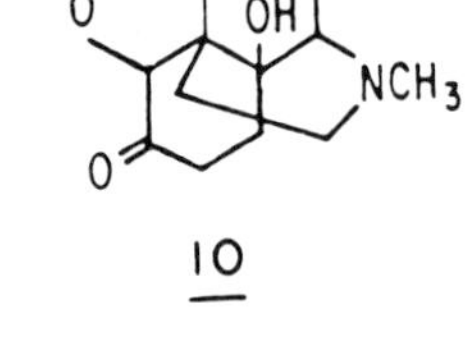

10

spectrometric procedures are developed for the identification of these compounds and the associated synthesis by-products and impurities.[64-72] Earlier studies[64] used conventional extraction-based procedures for the purification and isolation of these compounds prior to the use of MS (EI) for identification. Packed[65,66] and capillary[67,68] column GCMS are now preferred. Major impurities found in association with the Leuckart synthesis of methamphetmine[64-68] are shown as follows: **11** (*N*-methylformamide), **12** (methylbenzyl ketone), **13** (*N,N*-dimethylamphetamine), **14** (*N*-formylam-

Table 3
NUMBER OF ACTIVE HYDROGEN AND RELATIVE INTENSITIES OF PROTONATED AND ADDUCT IONS OF MORPHINE ALKALOIDS

		1	2	3	4	5	6	7	8	9	10[a]
Mol wt		285	285	285	299	299	311	311	271	297	315
Total no. of active H		2	2	1	1	0	0	1	3	0	1
No. of ally OH		1	1	0	1	0	0	0	1	0	1
No. of phenolic OH		1	0	1	0	0	0	1	1	0	0
Observed m/z of "protonated" ion											
NH$_3$		286	286	286	300	300	312	312	272	298	316
ND$_3$		289	289	288	302	301	313	314	276	299	318
Reagent gas[b]	Ion										
CH$_4$	(M + C$_2$H$_5$)$^+$, D	7.2	8.3	14	8.1	13	8.9	16	5.3	13	13
	(M + C$_2$H$_5$ – 18)$^+$, C	4.3	—[c]	—	—	—	—	—	2.7	—	1.0
	(M + H)$^+$, B	31	35	100	28	100	100	100	28	100	100
	(M + H – 18)$^+$, A	100	100	3.3	100	—	—	2.6	100	—	4.9
	(C + D)[d]	8.8	6.1	14	6.3	13	8.9	16	6.3	13	13
	(A + B)$^+$	100	100	100	100	100	100	100	100	100	100
NH$_3$	(M + NH$_4$)$^+$	81	77	100	98	100	—	—	32	98	—
	(M + H)$^+$	100	100	34	100	32	100	100	100	100	100

[a] See text and Structures 1-10 for the names and structures of these compounds.

[b] Data obtained by using isobutane as the reagent gas are parallel to those obtained under methane conditions and not presented or discussed.

[c] Less than 1.0% of the base peak.

[d] This represents a better estimate of the adduct ion to protonated ion intensity ratio.

From Liu, R. H., Low, I. A., Smith, F. P., Piotrowski, E. G., and Hsu, A. F., *Org. Mass Spectrom.*, 20, 511, 1985. With permission.

CH$_3$NHCHO   ◎-CH$_2$COCH$_3$ ◎-CH$_2$CHCH$_3$
                                                    |
                                                  N-CH$_3$
                                                    |
                                                  CH$_3$

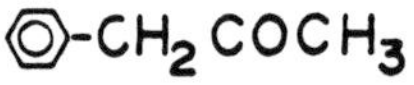

  $\underline{11}$ $\underline{12}$ 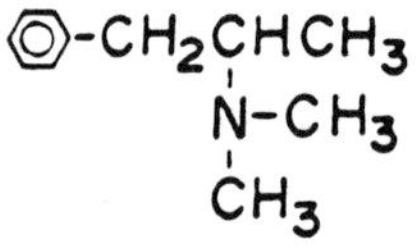 $\underline{13}$

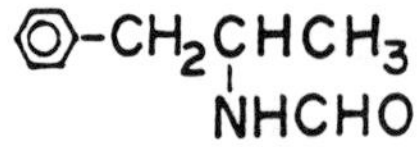

◎-CH$_2$CHCH$_3$ ◎-CH$_2$CHCH$_3$ ◎-CH$_2$\
        | | CO
      NHCHO   CH$_3$-NCHO   ◎-CH$_2$/

  $\underline{14}$ 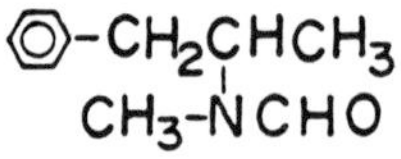 $\underline{15}$ 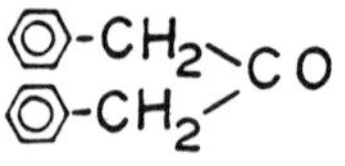 $\underline{16}$

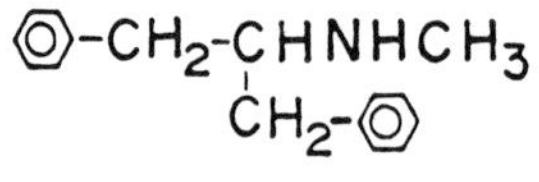

◎-CH$_2$-CHNHCH$_3$ ◎-CH$_2$CH$_2$-N-CH$_2$CH$_2$-◎
         | |
        CH$_2$-◎   CH$_3$

        $\underline{17}$ 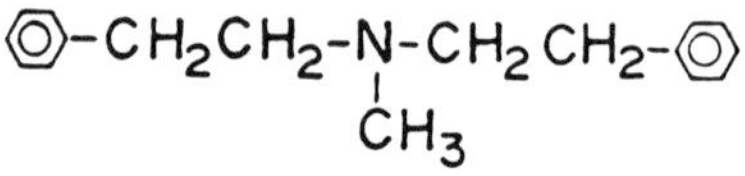 $\underline{18}$

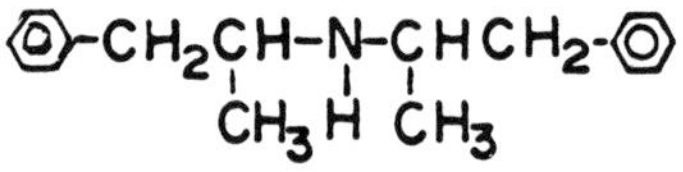

◎-CH$_2$CH-N-CHCH$_2$-◎   ◎-CH$_2$CH-N-CH-CH$_2$-◎
        | | | | | |
       CH$_3$ H CH$_3$ CH$_3$ CH$_3$ CH$_3$

          $\underline{19}$ 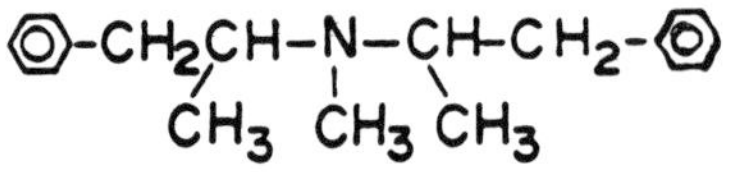 $\underline{20}$

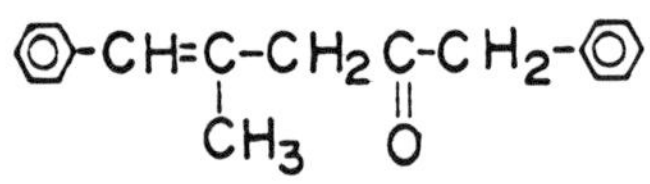

                                         CH$_3$ O
                                           | ‖
◎-CH=C-CH$_2$C-CH$_2$-◎   ◎-CH$_2$C=C-C-CH$_3$
       | ‖ |
      CH$_3$ O   ◎

       $\underline{21}$ 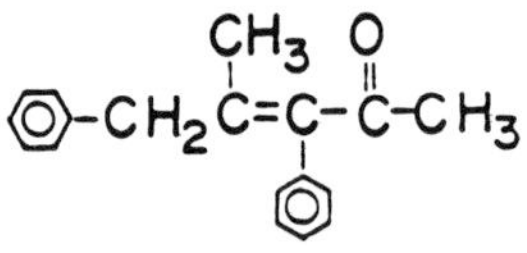 $\underline{22}$

                       H
                       |
        ◎-CH$_2$C=C-C-CH$_2$-◎
               | ‖
              CH$_3$ O

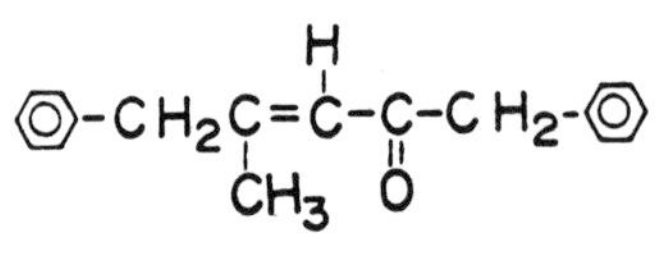

                $\underline{23}$

phetamine), **15** (*N*-formylmethamphetamine), **16** (dibenzylketone), **17** (*α*-benzyl-*N*-methylphenethylamine), **18** (*N*-methyldiphenethylamine), **19** (*α,α'*-dimethyldiphenethylamine diastereoisomer pair), **20** (*N,α,α'*-trimethyldiphenethylamine), **21** (1,5-diphenyl-4-methyl-4-penten-2-one), **22** (3,5-diphenyl-4-methyl-3-penten-2-one), and **23** (1,5-diphenyl-4-methyl-3-penten-2-one). It should be noted that some of these compounds may be artifacts produced during the analysis process.

Due to their possible psychotomimetic properties,[73] compounds with substitution of

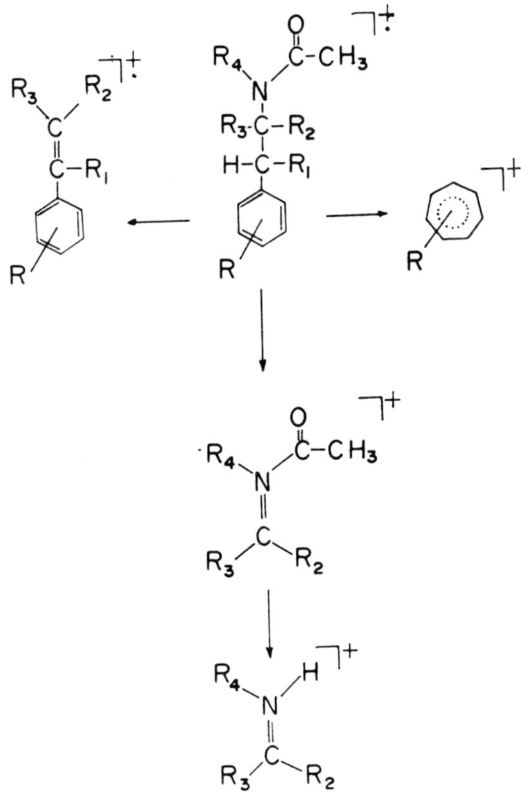

FIGURE 3. Major EI fragmentation pathways of amphetamine-related compounds. (From Coutts, R. T., Jones, G. R., Benderly, A., and Mak, A. L. C., *J. Chromatogr. Sci.*, 17, 350, 1979. With permission.)

methoxy groups on the benzene ring of amphetamine and related compounds are encountered in forensic laboratories.[74] The identification and differentiation of these compounds has been reported.[75-77] Tryptamines represent another category of amine drugs of interest to forensic chemists. The identification of major impurities in illicit tryptamine synthesis[78] has been addressed. Other amine drugs or related compounds often encountered in forensic laboratories include phenyl-2-propanone,[79] a primary precursor for amphetamine and methamphetamine manufacture, side chain positional isomers of amphetamine,[80] psilocybe,[81-83] mescaline,[84] and ephedrines.[84-86] The combination of GCMS and derivatization techniques are used for the separation and identification of enantiomeric ephedrine[84-86] and other related drug enantiomers.[84,85]

Mass spectrometric procedures are used in all of the studies cited above. Fragmentation of these compounds under EI are emphasized in several references[85-93] and found to follow a similar pattern as summarized in Figure 3.[88,90] The general fragmentation pattern of these compounds is the "β-fission" process (β to the amino function and to the aromatic system) with charge retained predominantly on the side chain as demonstrated by the methamphetamine spectrum (Figure 4). When the benzene ring is substituted by a methoxy group, the tendency of charge retention on the aromatic fragment increases.[84,87,88] Although substitution at different positions may alter the spectra to some extent, identification of these isomers based on mass spectra alone is difficult.[76,77,89] The presence of a hydroxyl group, as in ephedrine, also appears to enhance the retention of the charge on the aromatic fragment.[84,89-91]

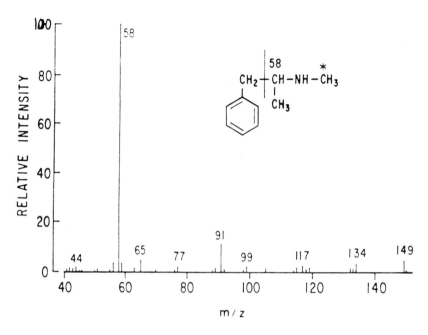

FIGURE 4. EI spectrum of methamphetamine showing the "β-fission" process.

Results obtained from CI analysis[86,90,94] of these compounds indicate that $(M + 1)^+$ ions are, in most cases, desirably the base peaks when isobutane is used as the reagent gas. Although the $(M + 1)^+$ ions of ephedrine and norephedrine isomers are not always the base peaks, their intensities are substantially higher compared to the corresponding M^+ ions obtained under EI conditions.

NCI has also been applied to the analysis of ephedrine (reagent gas, CO_2),[27] amphetamine (reagent gas, methane),[24] and amphetamine congeners (reagent gas, ammonia).[94] The CO_2 NCI spectrum shows the M^- ion as the base peak. Derivatized with pentafluorobenzoyl chloride or tetrafluorophthaloyl anhydride, amphetamine at the attomole (10^{-18}) level can be detected with NCI GCMS selected ion monitoring techniques. M^- ion is the base peak in this case. Ammonia NCI produces base peaks at $(M - 1)^-$, and small peaks which correspond to the loss of two hydrogens from the side chains.[94]

Many of the MS studies mentioned above are conducted in conjunction with GC. The incorporation of KOH (2%) to Apiezon L,[65,66] and Carbowax 20M[90] in packed columns is successfully used for the analysis of underivatized and derivatized amine drugs. When properly derivatized, these compounds are well separated by Carbowax 20M,[88] OV-101,[88] OV-1,[86] and OV-17[86] in packed column, and SP-2100 and Chirasil-Val (a chiral phase) in capillary column.[67,68] A mixture of nine amine drugs, 24 (amphetamine), 25 (methaphetamine), 26 (norephedrine), 27 (ephedrine), 28 (3,4-methylenedioxyamphetamine), 29 (mescaline), 30 (N,N-dimethyltryptamine), 31 (N,N-diethyltryptamine), 32 (N,N-dimethyl-5-methoxytryptamine), their possible enantiomers, and caffeine (33), are separated by a SE-54 fused silica glass capillary.[84] Most of these compounds are analyzed as N-trifluoroacetyl-1-prolyl derivatives (see 34 for the structure of the amphetamine derivative). The total ion and single ion current chromatograms are shown in Figure 5.[84] One is unlikely to encounter all of these compounds and enantiomers in one sample, and the separation of these compounds in a single chromatographic run is apparently beyond practical needs. However, with the establishment of the GC retention and mass spectrometric characteristics of these com-

Structure 24: phenyl-CH_2CHCH_3 with NH_2

Structure 25: phenyl-CH_2CHCH_3 with $NHCH_3$

Structure 26: phenyl-$CHCHCH_3$ with OH and NH_2

Structure 27: phenyl-$CHCHCH_3$ with OH and $NHCH_3$

Structure 28: methylenedioxyphenyl (H_2C-O, O) -CH_2CHCH_3 with NH_2

Structure 29: OCH_3, H_3CO, OCH_3 phenyl-$CH_2CH_2NH_2$

Structure 30: indole (H-N) -$CH_2CH_2N(CH_3)_2$

Structure 31: indole (H-N) -$CH_2CH_2N(C_2H_5)_2$

Structure 32: H_3CO-indole (H-N) -$CH_2CH_2N(CH_3)_2$

Structure 33: caffeine — CH_3-N, O, CH_3, N, O, N, CH_3, N

Structure 34: phenyl-CH_2CHCH_3, $N-C$, H, O, pyrrolidine ring N, $C=O$, CF_3

pounds in a single GCMS set of conditions, the identification of one or several amine drugs in a sample can be routinely performed. Finally, it is interesting to note that GCMS CI procedures can be used for the detection of amphetamine and methamphetamine in a single human hair.[95]

C. Marijuana and Other Cannabinoid-Containing Materials

Cannabinoid-containing samples, in the forms of marijuana (*Cannabis sativa* L.), hashish, and hashish oil, are commonly encountered in forensic laboratories. Although the use of microscopic examination, modified Duquenois-Levine color test, and chromatographic examination of extracts are usually sufficient for proving the sample under examination is indeed marijuana, GCMS, preferably capillary column,[96,97] is probably needed to satisfy a critical chemist. GCMS procedures have been applied to the analysis of traces of marijuana in hand swabs[98] and the identification of cannabinoids in pyrolytic products[99,100] and smoke condensates.[96,97,101]

An ideal method for the analysis of cannabinoid-containing samples in forensic laboratories should require minimum sample preparation, yet be specific enough to make a definite conclusion. Mass spectrometric procedures[102,103] are developed to meet these goals. For this approach, a minute amount (around 0.2 mg) of the pulverized raw sample is introduced into the mass spectrometer through a direct inlet probe. Intensities of selected ions (m/z 314, 310, 299, 295, 271, 258, 246, 243, 238, 231, and 193)

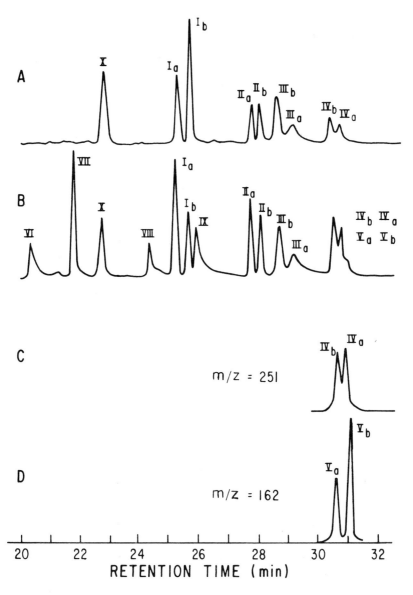

FIGURE 5.   Total ion chromatograms of (A) a mixture containing caffeine (X), *N*-TFA-*l*-prolyl-*l*-amphetamine (Ia), *N*-TFA-*l*-prolyl-*d*-amphetamine (Ib), *N*-TFA-*l*-prolyl-*d*-methamphetamine (IIa), *N*-TFA-*l*-prolyl-*d*-methamphetamine (IIb), *N*-TFA-*l*-prolyl-*d*-norephedrine (IIIb), *N*-TFA-*l*-prolyl-*l*-norephedrine (IIIa), *N*-TFA-*l*-prolyl-*d*-ephedrine (IVb), and *N*-TFA-*l*-prolyl-*l*-ephedrine (IVa); (B) a mixture containing the compounds in (A) plus mescaline (VI), *N,N*-dimethyltryptamine (VII), *N,N*-diethyltryptamine (VIII), *N,N*-dimethyl-5-methoxytryptamine (IX), *N*-TFA-*l*-prolyl-*l*-3,4-methylenedioxyamphetamine (Va), and *N*-TFA-*l*-prolyl-*d*-3,4-methylenedioxyamphetamine (Vb). Single ion current chromatograms of (C) compounds IVb, and IVa (m/z 251); (D) compounds Va, and Vb (m/z 162). (From Liu, R. J. H., Ku, W. W., and Fitzgerald, M. P., *J. Assoc. Off. Anal. Chem.*, 66, 1443, 1983. With permission.)

obtained from the sample under examination are related to the intensities of these ions obtained from standard compounds as follows:

$$Y_1 = X_{11}r_1C_1 + X_{12}r_2C_2 + \cdots + X_{1m}r_mC_m$$
$$Y_2 = X_{21}r_1C_1 + X_{22}r_2C_2 + \cdots + X_{2m}r_mC_m$$

$$\cdot$$
$$\cdot$$
$$\cdot$$

$$Y_n = X_1r_1C_1 + X_{n2}r_2C_2 + \cdots + X_{nm}r_mC_m \tag{1}$$

where Y_n is the observed relative intensity of the ion, n, in the sample; X_{nm} is the relative intensity of the ion, n, obtained from the standard cannabinoid, m; r_m is the relative sensitivity factor of the cannabinoid, m, under the experimental conditions; C_m is the concentration of the cannabinoid, m, in the sample under examination. An established statistical package, SPSS,[104] is used to perform the regression analysis with Xs and Ys obtained from MS experiments. In theory, data obtained from an unlimited number of compounds, m, can be fed into Equation 1 for regression analysis. In these cited studies, m is limited to the four major cannabinoids, Δ-1-tetrahydrocannabinol (Δ-1-THC), Δ-6-tetrahydrocannabinol (Δ-6-THC), cannabinol (CBN), and cannabidiol (CBD). The chemical structure, molecular weight, and possible interconversion of major cannabinoids are shown in Figure 6.

The presence of ions common to those found in standard cannabinoids indicate, qualitatively, that the sample under examination may indeed contain cannabinoids. Since samples normally contain different proportions of several cannabinoids which often generate common ions, the conventional approach in comparing the relative intensities of these ions cannot offer a definite conclusion for the presence of a specific cannabinoid. Using the described regression analysis approach, a high coefficient of determination, normally higher than 97%, can be considered as a quantitative indication of the certainty of the qualitative analysis. The conclusiveness of the qualitative analysis is further affirmed by the observation of high coefficients of determination with MS data obtained under several different levels of ionization energy, ranging from 14 to 20 eV. This represents a quantitative application of the semiquantitative "electron voltage mass fragmentation" approach[105-112] which will be described further in a later paragraph.

The regression coefficient (r_mC_m terms in Equation 1) obtained, by the regression analysis, for each compound indicates the relative concentration of this compound in the sample. It is further possible to use this parameter for differentiation.[103] Since standardization is not done in these studies,[106,107] the concentration term, C_m, cannot be extracted from the regression coefficient term, r_mC_m. (This is done in a later study applied to the analysis of polychlorinated biphenyl formulations, Aroclors.[113])

MS is used extensively in the structural characterization of compounds derived from *Cannabis sativa* L. *Cannabis* has been shown to contain over 400 compounds, of which 61 are known to be cannabinoids.[109] Since many of these cannabinoids differ only in the position of the double bond or the number of carbons in the side chain, and many have the same molecular weights, mass spectrometric characterization of these compounds are examined under several ionization energy levels, ranging from 5.5 to 21 eV.[102-112] "It is assumed that variation of electron ionization energy will reflect certain characteristic aspects of the stability of the parent compound. The rate at which a certain fragment may originate from different molecular ions will depend upon the activation energy required for fragmentation and upon the pathway available for the fragmentation."[112] Thus, a graph obtained by plotting the relative intensity of a se-

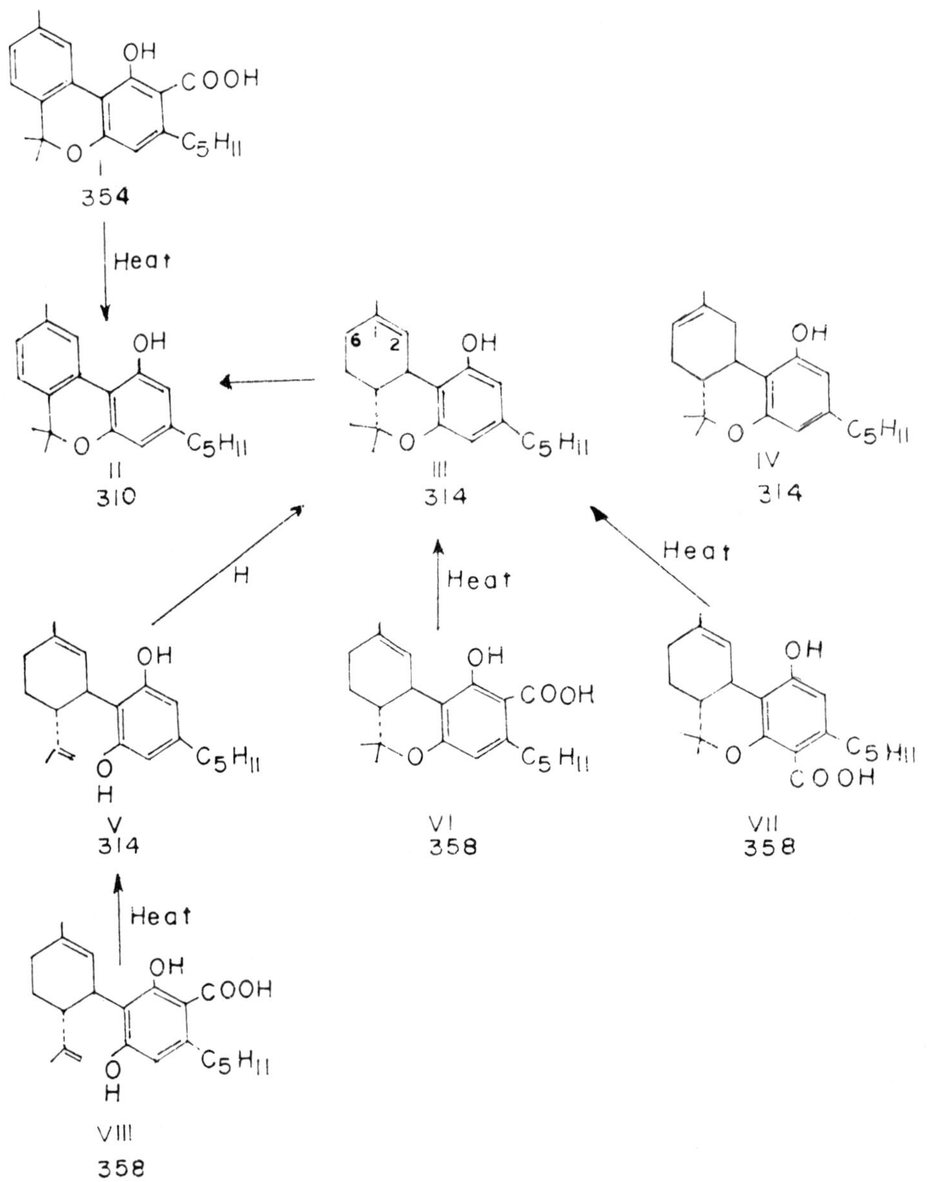

FIGURE 6.   Chemical structure, molecular weight, and possible interconversions of cannabinoic acid (I), cannabinol (II), Δ-1-THC (III), Δ-6-THC (IV), cannabidiol (V), Δ-1-THC acid A (VI), Δ-6-THC acid (VII), and cannabidiolic acid (VIII). (From Liu, R. J. H., Fitzgerald, M. P., and Smith, G. V., *Anal. Chem.*, 51, 1875, 1979. With permission.)

lected ion vs. the ionization electron voltage applied often reveals the characteristic features of a specific compound, and provides valuable information for compound differentiation.

Compounds shown in Figure 6 represent the major classes of cannabinoids of general interest. Basically, CBNs contain a pyran ring with four consecutive carbons as parts of a toluene and an alkyl phenol system. The toluene system is replaced by a monocyclic terpene moiety in CBDs and THCs. In CBDs the ether linkage in the pyran ring opens to form a hydroxyl group. Many of these three classes of cannabinoids also

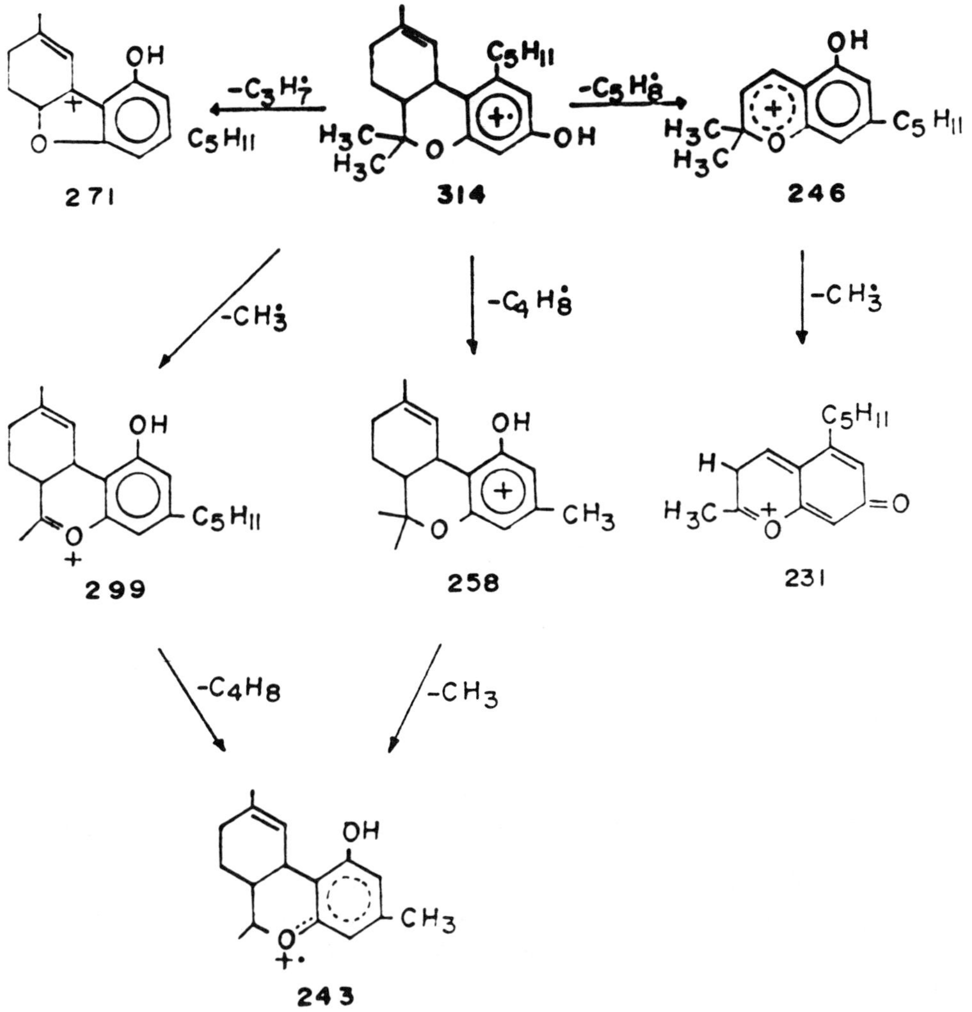

FIGURE 7. Major EI fragmentation pathways of Δ-1-THC. (From Vree, T. B., *J. Pharm. Sci.*, 66, 1444, 1977. With permission.)

exist with different number of carbons, mainly odd numbered, in the side chain,[105-107,115] and/or with one or more carboxyl groups. The position of the double bond in the terpene moiety may also vary. The phenolic group may also be replaced by an alkoxy group.[108] Many cannabinoids derived from other variations of these basic structural features have also been identified.[116,117]

EI fragmentation patterns of major cannabinoids have long been studied.[118-121] Fragmentation occurs preferentially in the alicyclic portion of the molecule, while the benzene ring functions as a charge-stabilizing element.[118] The most important fragmentation processes start with a retro Diels-Alder reaction of the cyclohexene ring.[119] Based on these earlier studies, and the effects of the ionization energy on the intensities of major ions (and their analogs in derivatives), the mechanisms for the formation of commonly observed higher mass ions, m/z 314, 299, 271, 258, 246, and 231, have been described in great detail.[109] Using Δ-1-THC as an example, the major fragmentation pathways are summarized in Figure 7.

Δ-1-THC, the principal active component of the plant material, has been analyzed by methane and methane-ammonia CI method.[122] Under a mixture of 2 × 10⁻⁴ mm

R₃, Y₁, N, 3, 4, 2, Y₂, R₁, 5, 6, N, H, R₂, O

35

ammonia and 8×10^{-4} mm methane, trimethylsilylated Δ-1-THC shows a single (M + 1)⁺ ion. The pentafluorobenzoyl derivative of Δ-1-THC in a 20-fg (10^{-15} g) level can be detected by NCI using methane as the reagent gas.[24] The intensity of the most abundant ion in NCI is reported to be more than 300 times higher than that observed under positive CI condition.

Many GCMS analyses of cannabinoids adopt prior chemical derivatization of these compounds. Major derivatization techniques will be summarized in Section IV.

D. Barbiturates

The ready availability of barbiturates, through both prescription and illicit sources, necessitates the development of reliable procedures for the analysis of these compounds. Molecular weights and the substituents in the ring system (35) of common barbiturates are listed in Table 4.

EI spectra of barbiturates[123-135] and their N-derivatized counterparts[127-135] follow similar patterns. Major EI fragmentation pathways, as depicted for *N,N'*-dimethyl derivatives, are summarized in Figure 8.[128] EI spectra of these compounds show the following characteristics:[125] (1) with the exception of sulfur analogs, molecular ions are absent or have low intensities;[129] (2) the ring system is stable; only when R_1 or R_2 is aromatic does cleavage of the ring system occur; (3) the loss of the longer saturated alkyl side chain from the molecular ion is preferred; thus, spectra obtained from butabarbital, butethal, pentobarbital, and amobarbital, and secobarbital, itobarbital, and butalbital show intragroup similarities; (4) when R_1 is an ethyl group and R_2 is not α-branched, route a (see Figure 8) becomes more significant; thus, the differentiation between amobarbital and pentobarbital, and butethal and butabarbital are possible.

With these basic fragmentation patterns, it becomes apparent that spectra obtained from aprobarbital, talbutal, and secobarbital are difficult to differentiate. This indeed has been reported in the case of aprobarbital and secobarbital.[136] Methane CI studies[136,137] are thus conducted to avoid this deficiency. Partially due to the large cross-section for proton capture possessed by the barbituric acid nucleus, protonated ions derived from these compounds are generally intense and useful for revealing molecular weight information. The CI approach, however, fails to differentiate isomeric barbiturates such as amobarbital and pentobarbital.[136] The combined use of EI and CI is therefore needed for successful identification of all barbiturates. With this in mind, argon-water CI studies on amobarbital and pentobarbital have been conducted,[137] and found to produce abundant protonated and fragment ions. The intensities of the latter ions follow the characteristic EI patterns. With the molecular weight information and the EI characteristics, the differentiation of these two compounds becomes possible.[137]

Table 4
STRUCTURES AND MAJOR NCI IONS OF BARBITURATES

Name	Mol wt	Y_1	Y_2	R_3	R_1 mass	R_2 mass	Rel. int. of major NCI ions[139]			
							(M–H)	(M–R_1)	(M–R_2)	(M–R_1–R_2+H)
Phenyl methyl barbituric acid	219	O	O	H	Methyl 15	Phenyl 77	9	22	100	2
Barbital	184	O	O	H	Ethyl; 29	Ethyl; 29	82	—	100	12
Probarbital	198	O	O	H	Ethyl; 29	Isopropyl; 43	33	9	100	1
Butethal	212	O	O	H	Ethyl; 29	Butyl; 57	65	45	100	8
Butabarbital	212	O	O	H	Ethyl; 29	1-Methylpropyl; 57	14	5	100	2
Pentobarbital	226	O	O	H	Ethyl; 29	1-Methylbutyl; 71	9	13	100	3
Amobarbital	226	O	O	H	Ethyl; 29	3-Methylbutyl; 71	65	49	100	8
Vinbarbital	224	O	O	H	Ethyl; 29	1-Methyl-1-butenyl; 69	16	8	100	11
Cyclobarbital	236	O	O	H	Ethyl; 29	1-Cyclohexen-1-yl; 81	11	11	100	4
Heptabarbital	250	O	O	H	Ethyl; 29	1-Cyclohepten-1-yl; 95	17	12	100	14
Phenobarbital	232	O	O	H	Ethyl; 29	Phenyl; 77	6	43	100	7
Aprobarbital	210	O	O	H	Allyl; 41	2-Methylethyl; 43	1	100	6	5
Idobutal	224	O	O	H	Allyl; 41	Butyl; 57	0.8	100	2	3
Butalbital	224	O	O	H	Allyl; 41	2-Methylpropyl; 57	0.5	100	3	3
Talbutal	224	O	O	H	Allyl; 41	1-Methylpropyl; 57	1	100	13	6
Nealbarbital	238	O	O	H	Allyl; 41	2,2-Dimethylpropyl; 71	1	100	2	2
Secobarbital	238	O	O	H	Allyl; 41	1-Methylbutyl; 71	1	100	13	4
Allobarbital	208	O	O	H	Allyl; 41	Allyl; 41	0.2	100	—	39
Cyclopal	234	O	O	H	Allyl; 41	2-Cyclopenten-1-yl; 67	4	100	27	19
Ibomal	288	O	O	H	—CH$_2$—CBr=CH$_2$; 107	Isopropyl; 43	2.5	100	3	1
Sigmodal	316	O	O	H	—CH$_2$—CBr=CH$_2$; 107	1-Methylbutyl; 71	—a	100	3	17
Brallobarbital	286	O	O	H	—CH$_2$—CBr=CH$_2$; 107	Allyl; 41	—a	100	13	30
Hexobarbital	236	O	O	CH$_3$	Methyl; 15	1-Cyclohexen-1-yl; 81	35	18	100	22
Metharbital	198	O	O	CH$_3$	Ethyl; 29	Ethyl; 29	100	—	28	3
Mephobarbital	246	O	O	CH$_3$	Ethyl; 29	Phenyl; 77	11	41	100	11
Enallypropymal	224	O	O	CH$_3$	Allyl; 41	Isopropyl; 43	17	100	6	1
Methohexital	262	O	O	CH$_3$	Allyl; 41	1-Methyl-2-pentynyl; 81	6	15	100	15
Thiopental	242	O	S	H	Ethyl; 41	1-Methylbutyl; 71	13	4	100	18
Thialbarbital	264	O	S	H	Allyl; 41	2-Cyclohexen-1-yl; 81	10	13	100	32

a 0.0—0.2.

From Jones, L. V. and Whitehouse, M. J., *Biomed. Mass Spectrum.*, 8, 231, 1981. With permission.

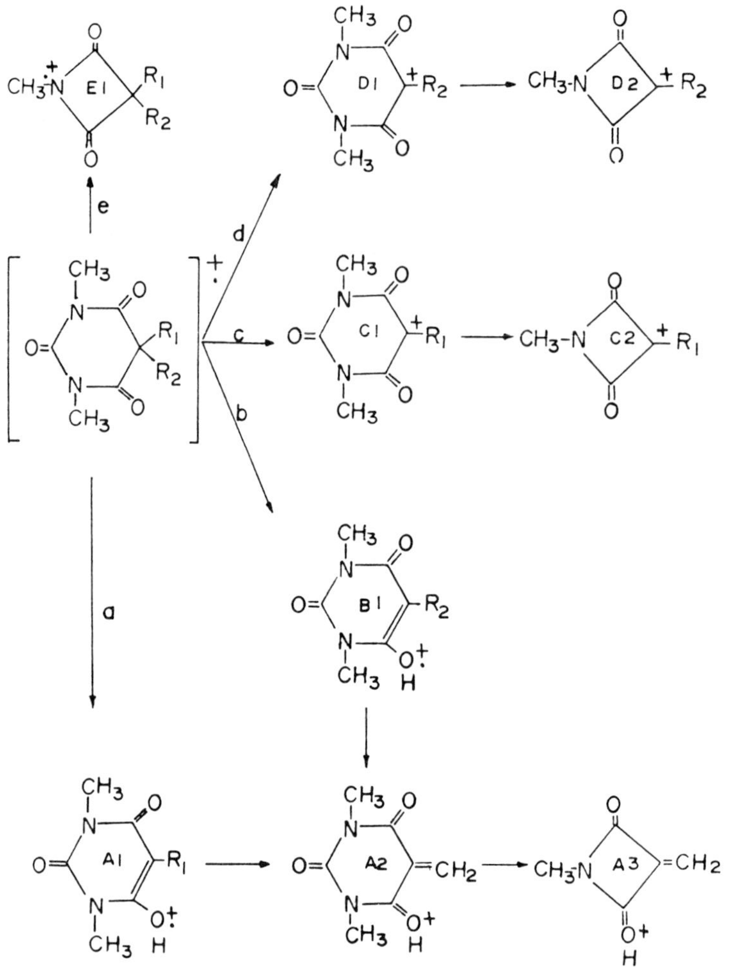

FIGURE 8.   Major EI fragmentation pathways of *N,N'*-dimethyl barbiturates. (From Skinner, R. F., Gallaher, E. G., and Predmore, D. B., *Anal. Chem.,* 45, 574, 1973. With permission.)

Since this approach has not been applied to the analysis of any extended list of barbiturates, whether it will produce a unique spectrum for each compound is not clear.

Reactant ion[21] and electron capture[138] NCI have also been applied to the analysis of barbiturates. Anions derived from the loss of side-chain radicals and a hydrogen atom provide valuable information concerning the molecular weight and side chain substituents. Perceiving this advantage, isobutane NCI studies of 30 underivatized barbiturates have been conducted.[139] By observing the intensities of $(M - H)^-$, $(M - R_2)^-$, and $(M - R_1 - R_2 + H)^-$ ions, all, but two, barbiturates are identified and differentiated. The relative intensities of these ions of the 30 compounds studied are listed in Table 4.

The following observations are generalized from this isobutane NCI study:[139] (1) NCI spectra from 5-allyl substituted barbiturates are relatively more intense than the corresponding 5-ethyl substituted ones; (2) *N,N'*-dimethylated barbiturates generate anion spectra of lower abundances compared to their parent compounds; (3) spectra from bromoallyl barbiturates are comparable with their nonbrominated analogs; (4) thiobarbiturates are characterized by base peaks corresponding to $(M - R_2)^-$ ions; (5) $(M - R_2)^-$ ions are the base peaks for 5-ethyl barbiturates, while $(M - R_1)^-$ ions are the most intense ones (with the exception of cyclopal) for 5-allyl compounds; (6) for 5-allyl compounds, branching at the carbon adjacent to the ring appears to enhance the $(M - R_2)^-$ ion. However, branching at the β-position does not appear to make any difference; spectra of idobutal and butalbital are virtually identical.

Structural similarity of this group of compounds makes them ideal examples for comparing EI, CI, and NCI spectral characteristics. The studies cited above strongly suggest the complementary nature of these techniques; only under rare and unusual occasions will one single ionization technique provide all characteristic information obtainable by every ionization method. Despite the claim that anion MS combines the advantages of EI and CI in providing structurally informative fragmentation patterns and molecular weight information, complete identification of all barbiturates, based on this technique alone, still cannot be done!

Retention characteristics of these compounds certainly could provide additional information for the differentiation of mass spectrometrically closely related barbiturates. The merits of a chromatographic component cannot be overemphasized. Although barbiturates themselves are amenable to chromatographing, N,N'-dimethyl derivatives provide improved results, and are commonly used in GCMS analysis.[127-132] SE-52,[127] OV-1,[128,129] XE-60,[130] SE-30,[131] and OV-101[132] have all been successfully used for this application.

E. Other Commonly Abused Drugs
1. Cocaine

At the present time, cocaine is probably one of the most abused controlled drugs. The unique nature of cocaine use in our society necessitates sophisticated analysis skill in handling this drug in forensic laboratories.[140] Various techniques[141-149] have been developed for the separation and identification of the following cocaine-related compounds: 36 (*d*-cocaine), 37 (*l*-cocaine), 38 (pseudococaine), 39 (allococaine), 40 (pseudoallococaine), 41 (3-aminomethyl-2-methoxycarbonyl-8-methyl-8-azabicyclo(3.2.1)oct-2-ene), 42 (3-benzoyloxy-2-methoxycarbonyl-8-methyl-8-azabicyclo(3.2.1)oct-2-ene), 43 (3-benzoyloxy-8-methyl-8-azabicyclo(3.2.1)oct-2-ene), 44 (cinnamoylcocaine), 45 (ecgonidine), and 46 (ecgonine).

EI spectra of cocaine-related compounds have been presented in several articles for sample identification purposes.[141-146] Fragmentation pathways have also been explored,[141,150,151] and are summarized in Figure 9. The establishment of these pathways is based on (1) the assumption that the charge of the molecular ion is initially localized on the nitrogen atom and, to a lesser extent, on the two carbonyl oxygen atoms;[150] and (2) observed mass shifts of ions derived from deuterated analogs[150,151] and related compounds, such as cinnamoylcocaine, benzoylecgonine, methylecgonine, and ethylecgonine.[141] High-resolution MS is also used to confirm the chemical formula proposed for specific ions.[151] Because of the olefinic bond at C_2, the mass spectrum of methylecgonidine differs somewhat[141] from the spectra of substituted ecgonines. The base peak at m/z 152 represents the formation of the substituted pyridine structure. Aromatization provides the driving force for this pathway. The spectrum of this compound, however, does contain the common characteristic ions at m/z 82 and 42.

Isobutane,[149] methane,[152,153] ammonia,[152] and methane-ammonia[153] CI spectra of these compounds have also been reported. The energetic protonated molecular ion derived from the methane CI conditions undergoes more fragmentation and produces ions of lower intensities.[152]

Since the identification of impurities in a cocaine sample provides valuable information, GCMS analysis is often applied. Stationary phases used include OV-1,[141,143,150-152,154] and OV-17.[144,151]

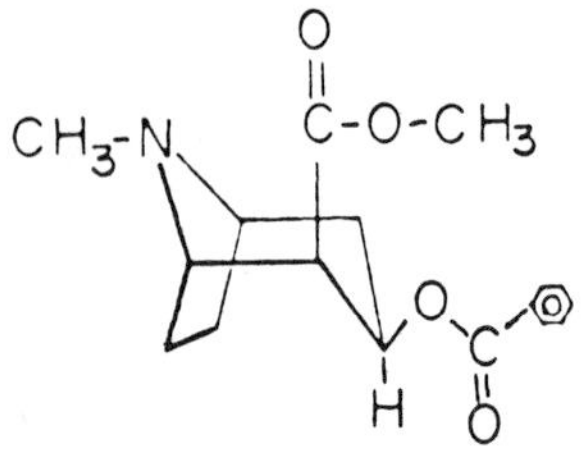

36 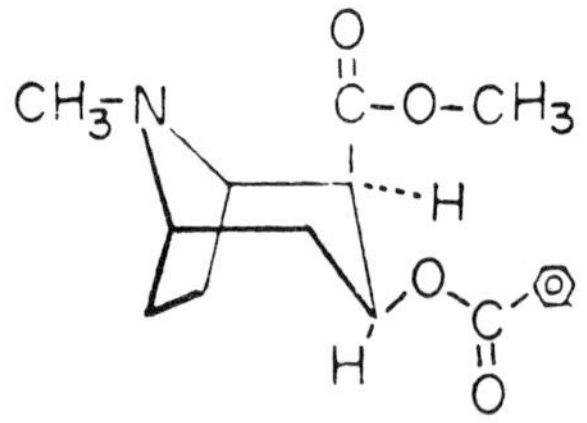   37 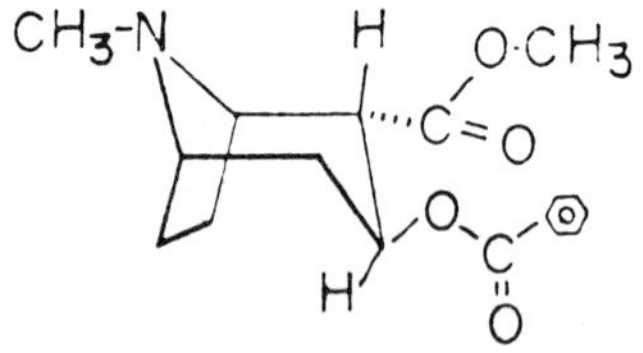   38

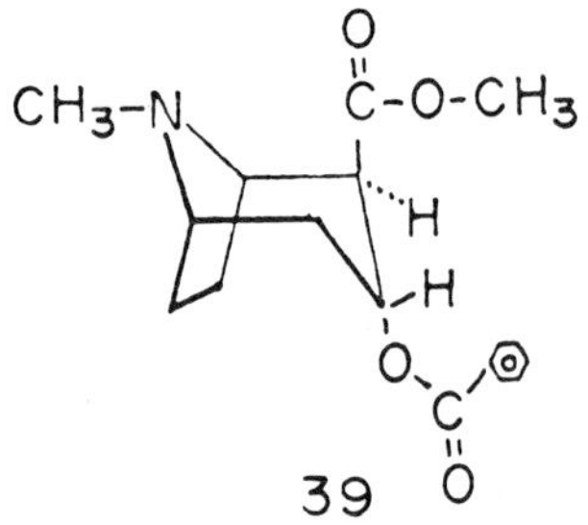

39 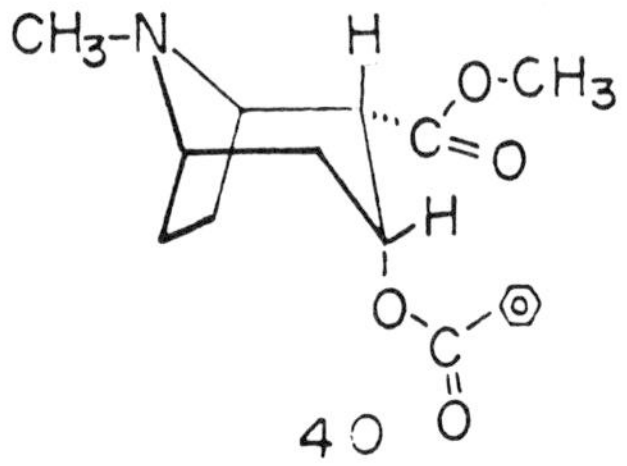   40 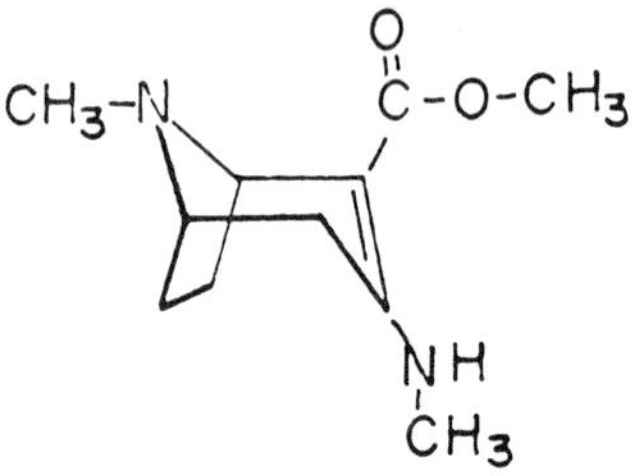   41

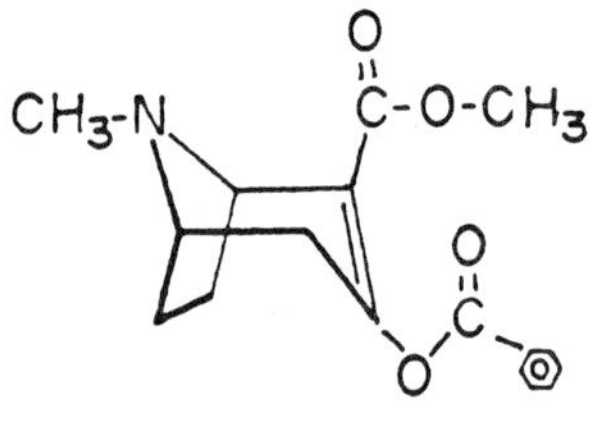

42 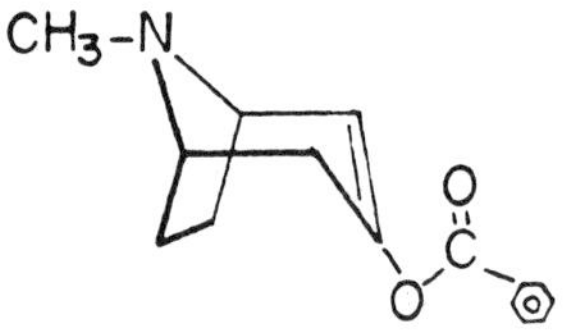   43 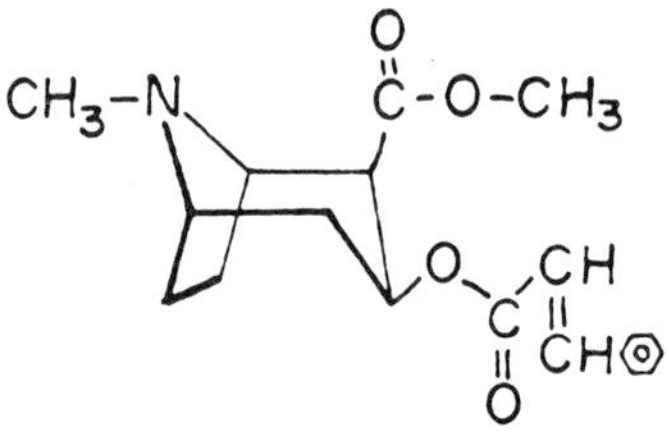   44

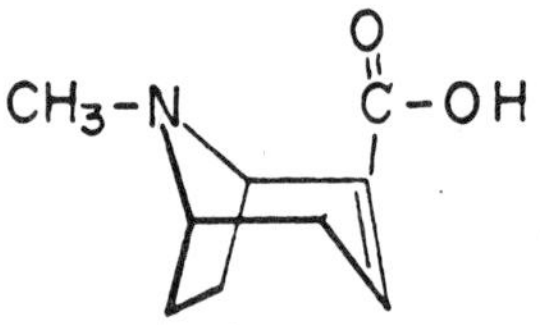

45 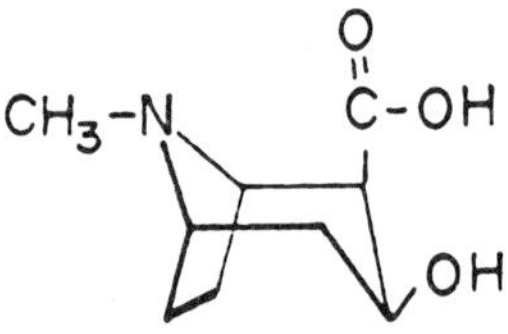   46

2. Phencyclidine and Analogs

When a particular drug is regulated, substances with similar structure and pharmacological effect are often introduced. The appearance of phencyclidine (PCP) analogs on the drug abuse scene reflects this characteristic of the drug abuse culture and many of these analogs[155-161] (Table 5 and 47 to 51) have actually been found in street samples.

EI spectra of these compounds have been systematically studied.[155,160,162,163] These spectra are very informative in revealing structural characteristics of these compounds. All compounds give M^+, with intensities ranging from 10 to 60%, $(M - 43)^+$ (loss of $-C_3H_7$ from the cyclohexyl ring), and to a lesser degree, $(M - 29)^+$ and $(M - 57)^+$ ions. Nitrogen heterocyclic compounds are characterized by the presence of the ions corre-

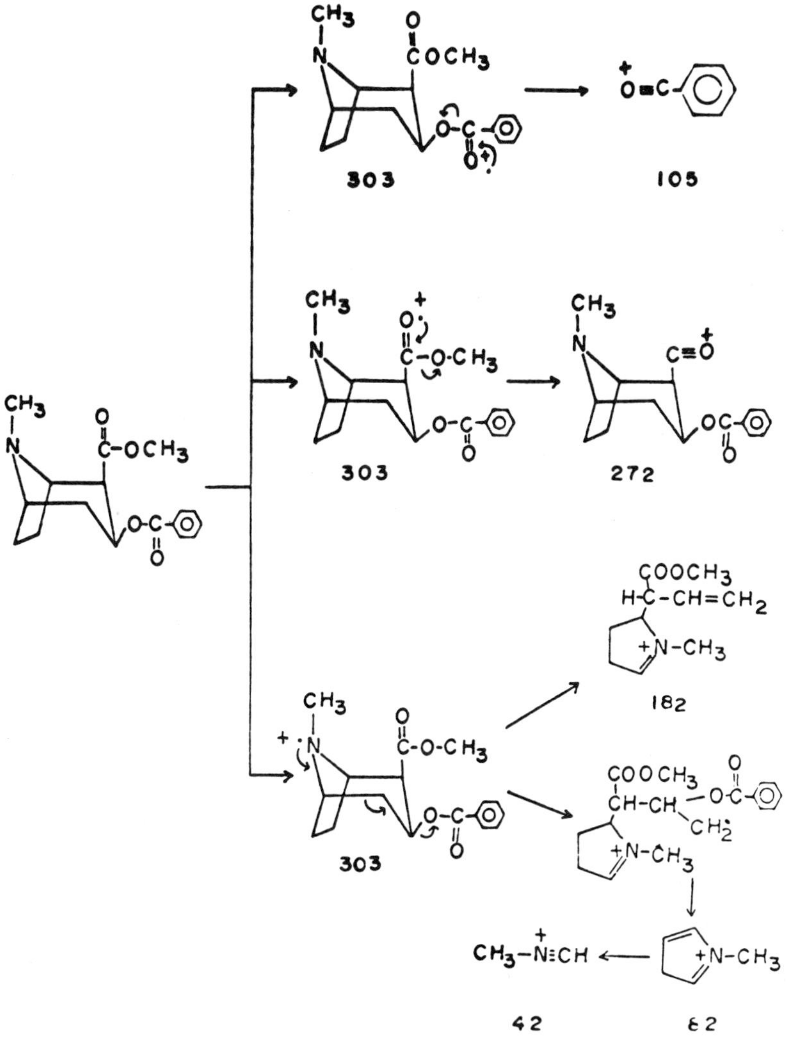

FIGURE 9. Major EI fragmentation pathways of cocaine. (From Jindal, S. P. and Vestergaard, P., *J. Pharm. Sci.*, 67, 811, 1978. With permission.)

sponding to the nitrogen heterocycle fragment and the complementary portion of the molecule. This phenomenon is even more pronounced in the thiophene-containing series. Thiophene and benzene compounds are characterized by the presence of $(C_5H_5S)^+$ and $(M - 83)^+$, and $(C_7H_7)^+$ and $(M - 77)^+$ ions.

Methane,[156,162-164] isobutane,[156] and methane-ammonia[164] CI have been used in combination with GC for the analysis of these compounds. Among the several phases (OV-7,[159] OV-207,[146,158,159,162] OV-25,[158] OV-225,[156,158] SP-2250,[160] and SE-30[156,158,163]) used, SE-30 appears to give the best overall separation of this series of compounds.[156]

3. Lysergic Acid and Related Compounds

The structures of various lysergic acid-related compounds are shown in 52 and Table 6.[92,165-167] The basic structure of the *d*-enantiomer is depicted in 52. Only one *iso*compound is shown in Table 6. The three-dimensional structures of *l*-enantiomers and *iso*compounds can be easily derived from the basic frameworks provided in 52 and Table 6. Among these compounds, *d*-LSD (first synthesized in 1943[168]) was reported to be the most effective hallucinogen.

Table 5

STRUCTURES AND MAJOR EI IONS OF PCP-RELATED COMPOUNDS[a]

Compound name (and designation)	Mol wt	R_1 (mass)	R_2 (mass)[b]	Characteristic EI ions[c,d]
1-(1-Phenylcyclohexyl)piperidine (PCP)	243	C_6H_6 (77)	Pi (84)	M-R₁; C₇H₇; R₁; M-R₂
1-(1-Phenylcyclohexyl)-4-methylpiperidine (PCMeP)	257	C_6H_6 (77)	4-CH_3-Pi (98)	M-R₁; C₇H₇; R₂; M-R₂
1-(1-Phenylcyclohexyl)-4-hydroxypiperidine (PCHP)	259	C_6H_6 (77)	4-HO-Pi (100)	M-R₁; C₇H₇; R₂; M-R₂
1-(1-Phenylcyclohexyl)pyrrolidine (PCPY)	229	C_6H_6 (77)	Py (70)	M-R₁; C₇H₇; R₂; M-R₂
1-(1-Phenylcyclohexyl)morpholine (PCM)	245	C_6H_6 (77)	M (86)	M-R₁; C₇H₇; R₂; M-R₂
N-Ethyl-1-phenylcyclohexylamine (PCE)	203	C_6H_6 (77)	NHC_2H_5 (64)	M-R₁; C₇H₇
N,N-Diethyl-1-phenylcyclohexylamine (PCDE)	231	C_6H_6 (77)	$N(C_2H_5)_2$ (112)	M-R₁; C₇H₇
N-Methyl-1-phenylcyclohexylamine (PCMe)	189	C_6H_6 (77)	$NHCH_3$ (30)	M-R₁; C₇H₇
N,N-Dimethyl-1-phenylcyclohexylamine (PCDMe)	203	C_6H_6 (77)	$N(CH_3)_2$ (44)	M-R₁; C₇H₇
N-Propyl-1-phenylcyclohexylamine (PCPr)	217	C_6H_6 (77)	NHC_3H_7 (58)	M-R₁; C₇H₇
N-Isopropyl-phenylcyclohexylamine (PCiPr)	217	C_6H_6 (77)	$NH\text{-}i\text{-}C_3H_7$ (58)	M-R₁; C₇H₇
N-Butyl-phenylcyclohexylamine (PCBu)	231	C_6H_6 (77)	NHC_4H_9 (72)	M-R₁; C₇H₇
1-Piperidinocyclohexanecarbonitrile (PCC)	192	CN (26)	Pi (84)	M-R₂; R₂
1-Morpholinocyclohexanecarbonitrile (MCC)	194	CN (26)	M (86)	M-R₂; R₂
1-Pyrrolidinocyclohexanecarbonitrile (PYCC)	178	CN (26)	Py (70)	M-R₂; R₂
1-Diethylaminocyclohexanecarbonitrile (DEACC)	152	CN (26)	$N(C_2H_5)_2$ (112)	
1-[1-(2-Thienyl)cyclohexyl]piperidine (TCP)	249	2-TH (83)	Pi (84)	M-R₁; C₅H₅S; M-R₂; R₂
1-[1-(2-Thienyl)cyclohexyl]pyrrolidine (TCPy)	235	2-TH (83)	Py (70)	M-R₁; C₅H₅S; M-R₂; R₂
1-[1-(2-Thienyl)cyclohexyl]morpholine (TCM)	251	2-TH (83)	M (86)	M-R₁; C₅H₅S; M-R₂; R₂

[a] See *47* for the general structure of these compounds. Abstracted from References 155, 160, 162, and 163.

[b] Pi, Py, M, and 2-TH represent the structures of *48, 49, 50,* and *51*, respectively.

[c] Ions are not listed in any intensity order.

[d] m/z M⁺, (M-43)⁺, (M-29)⁺, and (M-57)⁺ are common to all compounds, but not listed in this column.

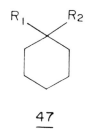

47

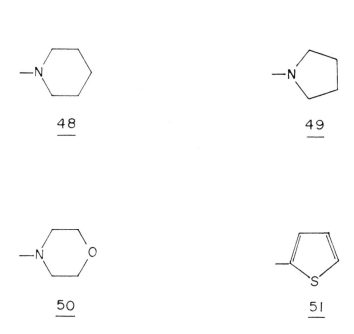

48

49

50

51

EI spectra of these compounds have been well characterized.[92,165-167,169-173] The basic fragmentation patterns are shown in Figure 10[167] as illustrated for lysergic acid. All compounds display intense molecular ions which nicely provide the needed molecular weight information. Characteristic pathways include (1) the loss of a hydrogen yielding a highly conjugated immonium ion in the D ring (route a); (2) the loss of $-C_2H_5N$ group (the 5-6 and 7-8 cleavages) derived from retro Diels-Alder reaction (route b); (3) the loss of $-C_3H_4O_2$ group (the 6-7 and 8-9 cleavage) (route c); and (4) various ways of eliminating the whole or part of the sidechain at the C_8 position (routes d, e, f, and g). Fragments produced by routes a, d, e, and f are subjected to further fragmentation as shown in the figure.

The relative significances of these routes are drastically influenced by the side chain at the C_8 position, the position of the double bond, and the substituent at the N_6 position. Intensities of ions derived from various fragmentation routes are available in the literature and summarized in Table 6. These data indicate the following fragmentation characteristics: (1) the relative significance of the retro Diels-Alder reaction is affected by the substituent at the N_6 position in the following order $CONH_2$ > H > CH_3 > CN, and it becomes inoperative when the double bond is shifted to the 8, 9 position; (2) routes b and c become insignificant when the substituent at the C_8 position is an alkyl or an alkyl-derived group; (3) route c is less favorable when the carboxylic group in the C_8 position is not derivatized; (4) saturation at the D-ring eliminates (or at least lessens) several fragmentation routes and produces a much more intense molec-

Table 6

STRUCTURES AND EI MS CHARACTERISTICS OF LYSERGIC ACID-RELATED COMPOUNDS

Compound name	Mol wt	Double bond	R_1	R_2	R_3	R_4	R_5	a	b	c	d	e	f	g	h	Ref.
d-Lysergic acid	268	9,10	H	H	CH_3	COOH	H	21	20	—	43	68	20	21	41	165-167, 169, 170
d-iso-Lysergic acid	268	9,10	H	H	CH_3	H	COOH									166, 170
d-Lysergic acid amide	267	9,10	H	H	CH_3	$CONH_2$	H									165
d-Lysergic acid dimethylamide	295	9,10	H	H	CH_3	$CON(CH_3)_2$	H									166
d-Lysergic acid ethylamide	295	9,10	H	H	CH_3	$CONH(C_2H_5)$	H	—	8	36	32	—	—	21	51	92,165,166
d-Lysergic acid diethylamide (LSD)	323	9,10	H	H	CH_3	$CON(C_2H_5)_2$	H	—	5	17	32	—	—	36	64	92
d-Lysergic acid methylpropylamide	323	9,10	H	H	CH_3	$CON(CH_3)C_3H_7$	H		7	61	9			11	15	165
d-Lysergic acid ethylpropylamide	337	9,10	H	H	CH_3	$CON(C_2H_5)C_3H_7$	H									166
d-Lysergic acid dipropylamide	351	9,10	H	H	CH_3	$CON(C_3H_7)_2$	H									166
2-Bromo d-LSD	401	9,10	H	Br	CH_3	$CON(C_2H_5)_2$	H	—	1	19	74	—	—	40	53	92
1-Methyl d-LSD	415	9,10	CH_3	H	CH_3	$CON(C_2H_5)_2$	H	—	3	21	38	—	—	32	52	92
1-Acetyl d-LSD	365	9,10	$COCH_3$	H	CH_3	$CON(C_2H_5)_2$	H	—	11	9	14	—	—	28	41	92
N^6-Demethyl d-LSD	309	9,10	H	H	H	$CON(C_2H_5)_2$	H	—	30	56	44	—	—	43	100	165, 171
N^6-Cyano-N^6-demethyl d-LSD	334	9,10	H	H	CN	$CON(C_2H_5)_2$	H	—	3	55	55	—	—	44	15	165, 171
N^6-Carbamoyl-N^6-demethyl d-LSD	352	9,10	H	H	$CONH_2$	$CON(C_2H_5)_2$	H	—	93	4	5	—	—	3	1	165, 171
Lysergol	254	9,10	H	H	CH_3	CH_2OH	H									165
1-Methylmethyl-ergonovine	253	9,10	CH_3	H	CH_3	$CONHCH\text{-}(C_2H_4OH)C_2H_5$	H	—	1	88	26	—	—	61	100	92
Argroclavine	238	8,9	H	H	CH_3	CH_3	—									165
Elymoclavine	254	8,9	H	H	CH_3	CH_2OH	—									165, 169
9,10-Dihydro LSD	325	None	H	H	CH_3	COOH	H	—	0.7	—	5	2	1	3	9	167

Rel. int. of ions from dif. routes (Figure 10)

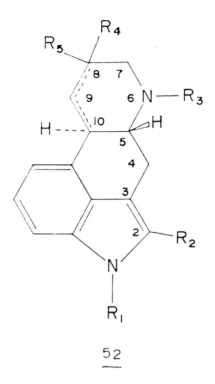

$\underline{52}$

ular ion which shows a mass shift of two with respect to its unsaturated counterpart. $(M - COOH)^-$ and the $(M - H)^-$ ions are the most significant peaks in the NCI spectra[167] of lysergic acid and its saturated counterpart.

4. α-Methylfentanyl

α-Methylfentanyl was first found in the illicit narcotic trade of southern California streets in 1981 under the name of "China White", a street term for very pure Southeast Asian heroin.[174] (The identification of this compound[175] by the Drug Enforcement Administration scientific team represents a modern version of a Sherlock Holmes episode. Those who want to enter the forensic science field should be prepared to equip themselves with the analytical skill demonstrated by these scientific investigators, but guard themselves against expecting magic solutions and thrills as depicted by the "Quincy" series.)

α-Methylfentanyl as shown in 53 has been reported[175] as a potent analgesic agent. The structure of this compound resembles that of fentanyl, 54, which has been well studied[174,176-178] and illicitly used for "doping" racehorses.[176] The EI spectrum of this compound is parallel to that reported for N-[1-(2-phenethyl-4-piperidinyl)] acetanilide,[178] 55, and thoroughly studied in a tandem MS (MSMS) investigation.[179] The secondary methyl group at the α-carbon greatly enhances the fragmentation pathways involving the loss of the benzyl group. Consequently, the molecular weight information is obtained through CI techniques.[175]

5. Methaqualone and Mecloqualone

The spectrometric and chromatographic characteristics of 16 substituted quinazolinones (56 and Table 7) have recently been reported.[180] The fragmentation patterns of this group of compounds are similar to those reported for silylated methaqualone metabolites.[181] All compounds display significant intensities of M^+ and $(M - 15)^+$ ions under 70 eV EI conditions. The relative intensities of characteristic ions, m/z 143 (see

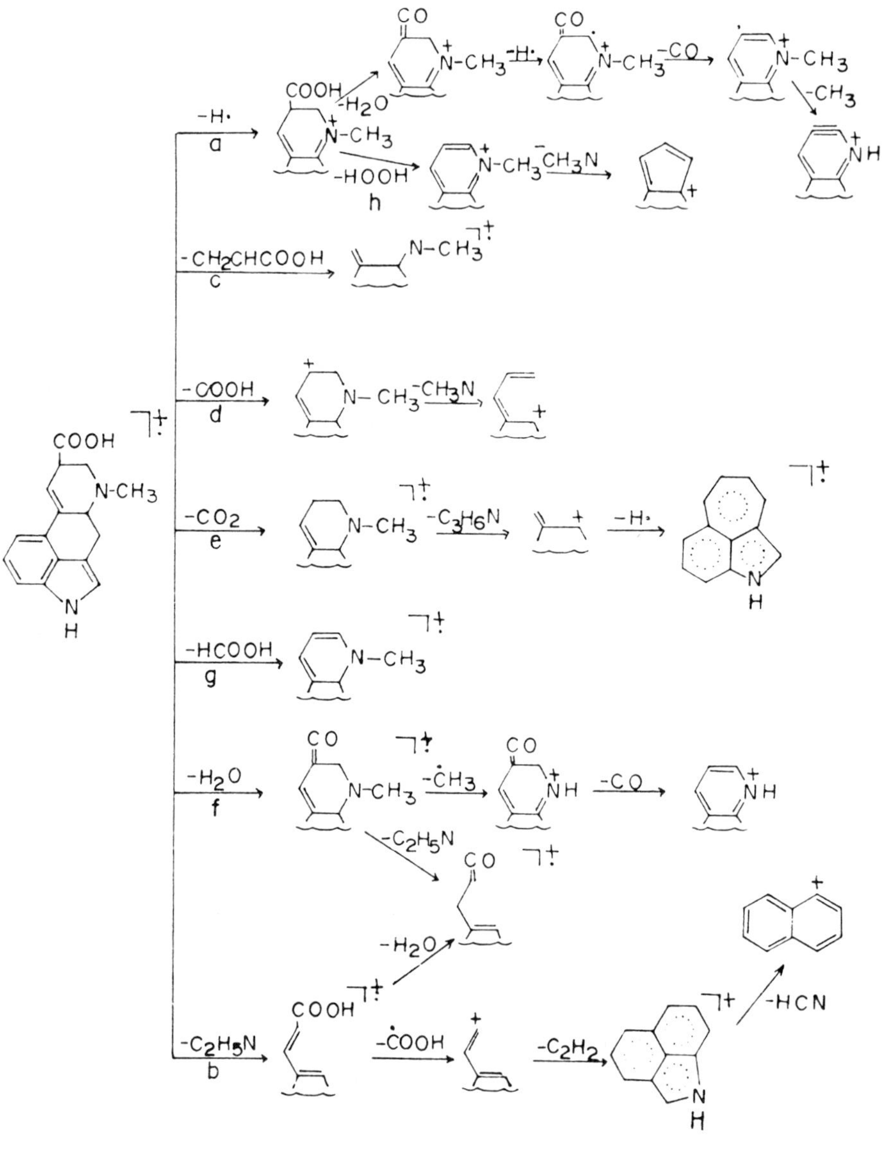

FIGURE 10.   Major EI fragmentation pathways of lysergic acid. (From Schmidt, J., Graft, R., and Voigt, D., *Biomed. Mass Spectrom.*, 5, 674, 1978. With permission.)

56), M$^+$, and (M − R)$^+$, R = F, Cl, Br or I, are measured and calculated from the line diagrams provided by the original publication[181] and listed in Table 7. The intensity ratios of these ions provide valuable information for the differentiation of these positional isomers.

IV. ANALYTICAL APPROACHES

A. Tandem Mass Spectrometry

Various forms of the developing MS/MS have been used for the analysis of drugs. These applications include Fourier transform MS (FT-MS),[182] mass-analyzed ion kinetic energy spectrometry (MIKES),[28-34] and triple-quadrupole MS.[183-186] A liquid chromatograph has also been coupled to a MSMS system as the sample introduction device.[41]

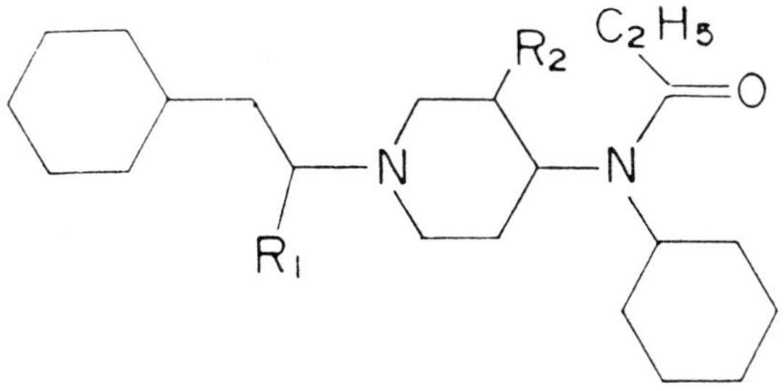

53 : $R_1 = CH_3$; $R_2 = H$

54 : $R_1 = R_2 = H$

55 : $R_1 = H$; $R_2 = CH_3$

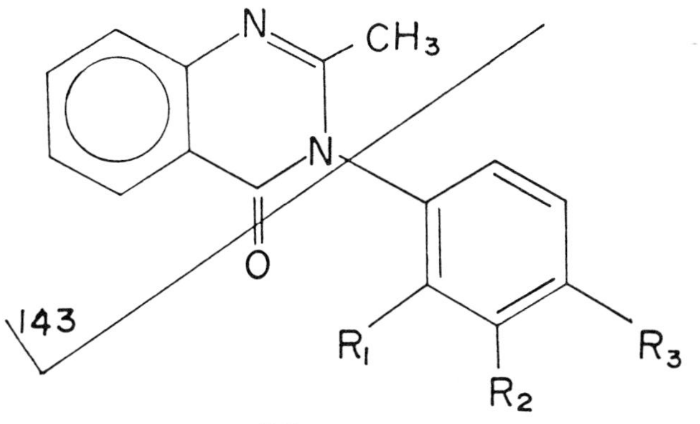

56

MSMS promises to provide valuable information for the identification of a compound of interest in a complex matrix, and for the elucidation of an unknown structure as well as the differentiation of structurally similar compounds. MIKES has been used for the mapping of the cocaine to cinnamoylcocaine ratio in whole coca plant tissues, without sample pretreatment or prefractionation.[32,33] The experiments employ direct probe vaporization, CI, and collision-induced dissociation. As the same is evaporated, both compounds are characterized by their protonated ions and the dissociation involving the loss of benzoic acid and cinnamic acid from the protonated molecules. (See 36 to 46 for the structure characteristics of these compounds.) Similar approaches are used to identify papaverine in raw opium,[31] and mescaline in crude plant extracts.[29,30]

Molecular structure determinations of "China White" and related fentanyl derivatives best illustrate the potential use of MSMS for structural elucidation.[179] The structure is pieced together through successive pairing of EI "complementary ions", whose sum of masses or elemental compositions equals that of the molecular ion, and the determination of the substructures of these pair ions by their collision-activated dissociation spectra. Spectra of the protonated or ethylated molecules generated by methane

Table 7

STRUCTURE AND EI MS CHARACTERISTICS OF METHAQUALONE COGENERS[a]

Compound name	R_1	R_2	R_3	Mol wt	Int.[b] of $(M-R)^+$	Int. ratios[b] $(M-R)^+:M^+:143$
2-Methyl-3-phenyl-4-quinazolinone	H	H	H	236	72	
2-Methyl-3-*o*-tolyl-4(3H)-quinazolinone	CH₃	H	H	250	100	1:0.86:0.33
2-Methyl-3-*m*-tolyl-4(3H)-quinazolinone	H	CH₃	H	250	93	1:0.70:0.030
2-Methyl-3-*p*-tolyl-4(3H)-quinazolinone	H	H	CH₃	250	81	1:0.96:0.97
2-Methyl-3-*o*-fluorophenyl-4-quinazolinone	F	H	H	254	100	1:0.90:0.54
2-Methyl-3-*m*-fluorophenyl-4-quinazolinone	H	F	H	254	21	1:4.5:2.5
2-Methyl-3-*p*-fluorophenyl-4-quinazolinone	H	H	F	254	8.3	1:11:7.2
2-Methyl-3-*o*-chlorophenyl-4-quinazolinone	Cl	H	H	270	100	1:0.19:0.33
2-Methyl-3-*m*-chlorophenyl-4-quinazolinone	H	Cl	H	270	57	1:1.5:1.1
2-Methyl-3-*p*-chlorophenyl-4-quinazolinone	H	H	Cl	270	38	1:2.6:2.3
2-Methyl-3-*o*-bromophenyl-4-quinazolinone	Br	H	H	315	100	1:0.056:0.083
2-Methyl-3-*m*-bromophenyl-4-quinazolinone	H	Br	H	315	86	1:0.29:0.53
2-Methyl-3-*p*-bromophenyl-4-quinazolinone	H	H	Br	315	25	1:1.8:3.0
2-Methyl-3-*o*-iodophenyl-4-quinazolinone	I	H	H	362	100	1:0.11:0.13
2-Methyl-3-*m*-iodophenyl-4-quinazolinone	H	I	H	362	43	1:1.3:0.77
2-Methyl-3-*p*-iodophenyl-4-quinazolinone	H	H	I	362	19	1:3.1:4.0

[a] See *56* for the positions of R_1, R_2, and R_3 groups.
[b] Intensity and intensity ratio information are measured and calculated from the spectra in Reference 180. R represents CH₃, F, Cl, Br, or I.

CI also provide complementary information not available from other ionization methods. This latter application is nicely illustrated in a study concerning the analysis of barbiturates as described in a previous section.[34]

B. Drug Sample Differentiation

"A major and overriding characteristic that sets criminalistics apart from other scientific disciplines is its unique concern with the process of individualization. Other sciences are satisfied when an object is classified into a unit place in the taxonomy of the discipline. Criminalistics strives to relate the object to a particularized source."[187] Drug sample individualizations,[188] or, more realistically, the determinations of common manufacture or trade origins, are specific examples of this general statement. MS-based methods are utilized in several approaches aiming for this analytical goal.

1. Qualitative and Quantitative Composition Determination

Sample differentiation based on the identification and quantitation of diluents, adulterants, impurities, and minor components are well illustrated for the studies of amphetamine,[64,65,189] cocaine,[141-145,190] morphine alkaloids,[191-194] and marijuana.[195] MS and GCMS methods are used in these studies for compound identification and quantitation.

The analysis of cocaine serves as a good example for the illustration of this point.[141-143] Defense attorneys, perhaps with the assistance of their scientific experts, have been reported to raise the natural vs. synthetic origin issue.[141] Cocaine derived from natural origin may come from either the extraction of coca leaves followed by successive recrystalizations, or the hydrolysis of all ecgonine-based alkaloids of the leaf with subsequent esterification. The presence of cinnamoylcocaine is often used as an indication of the former procedure, while the latter procedure, as a result of incomplete esterification of the ecgonine base, often produces ecgonine, methylecgonine, and benzoylecgonine. Thus, the presence of minor alkaloids of E. *coca, cis-* and *trans*-cinna-

Table 8
RELATIVE INTENSITIES OF SELECTED IONS OF CBN, CBN, Δ-1-THC, AND Δ-6-THC

Sample	314	310	299	295	271	258	246	243	238	231	193
CBN	0	13.0	7	100	0	0	0	0	30.9	0	0
CBD	8.2	0	1.3	0	0	1.1	22.8	0	0	100	13.7
Δ-1-THC	43.7	0	49.7	0	37.5	33.0	15.3	37.9	0	100	31.4
Δ-6-THC	41.5	0	5.6	0	27.1	33.9	17.9	4.9	0	100	25.3
Hashish oil	14.2	0.6	11.0	4.2	7.1	4.9	17.4	6.8	1.3	100	13.2

From Liu, R. J. H. and Fitzgerald, M. P., *J. Forensic Sci.,* 25, 815, 1980. With permission.

Table 9
COMPARISON OF MULTIPLE REGRESSION COEFFICIENTS OBTAINED FROM VARIOUS CANNABINOID-CONTAINING SAMPLES

Sample	CBN	CBD	Δ-1-THC	Δ-6-THC
Hashish oil	0.76	0.79	0.23	0
Hashish-1	0.68	0.91	0.10	0.032
Hashish-2	1.0	0.50	0.071	0
Hashish-3	1.0	0.20	0.35	0.28
Hashish-4	0.63	0.90	0.092	0.050
Leaf	0.50	0.17	0.99	0

From Liu, R. J. H. and Fitzgerald, M. P., *J. Forensic Sci.,* 25, 815, 1980. With permission.

moylcocaine, provides indirect proof of the natural origin.[143] Interpretation based on the presence of ecgonine, methylecgonine, benzoylecgonine, methylanhydroecgonine, and ethylbenzoylecgonine should be exercised with caution, as they may be decomposition products or artifacts generated during the analytical process.[141] On the other hand, cocaine derived from total synthesis, unless scrupulously purified, contains diastereoisomers and enantiomers.[144-146] The observation of impurities, such as 3-aminomethyl-2-methoxycarbonyl-8-methyl-8-azabicyclo(3.2.1)oct-2-ene, 3-benzoyloxy-2-methoxycarbony-8-methyl-8-azabicyclo(3.2.1)oct-2-ene, and 3-benzoyloxy-8-methyl-8-azabicyclo(3.2.1)oct-2-ene, further confirms the synthetic origin.

Quantitative distribution of major components may also constitute a basis for sample differentiation. The direct inlet probe MS/multiple regression analysis method described in the marijuana section[102,103] represents a mass spectrometric approach for the differentiation of multicomponent samples such as cannabinoid-containing samples.

In order to differentiate hashish oil, hashish, and marijuana leaf samples, these samples are introduced in the ion source of a mass spectrometer and spectra are continually collected until the sample is exhausted. The intensities of selected ions in these spectra from standard compounds and the actual samples are summed, respectively, and relative intensities calculated as shown in Table 8. Multiple regression analysis are performed according to the relationship set up in Equation 1. Regression coefficients, which reflect the relative concentrations of the respective cannabinoids, obtained from several samples are listed in Table 9. These results indicate that among the four hashish samples investigated, the relative concentrations of CBN, CBD, and Δ-1-THC in samples 2 and 3 are distinctly different from each other and from samples 1 and 4. However, samples 1 and 4 are similar.

The mass spectrometric aspect of this approach requires minimal effort. With the data system available to most mass spectrometric devices and the widespread availability of personal computers, the combined use of regression analysis and MS procedures adopted here appears to have a potential in routine analytical work.

2. Diastereoisomer and/or Enantiomer Analysis

Strictly speaking, diastereoisomer and enantiomers are different compounds; sample differentiation based on the variation of these components belongs to the same category treated under the preceding heading. Due to their structural similarities, but differences in biological activities and possible governmental regulation, the determination of these compounds are emphasized.

The possibility of differentiating a "synthetic" vs. a "natural" cocaine sample, based on the observation of other diastereoisomers has been addressed under the last heading. It is also interesting to note that diastereoisomers, quinine (a bitter flavoring agent for tonic water and drug adulterant) and quinidine (a cardiovascular agent), show distinct EI spectra.[196]

The possibility of misidentifying[197] *l*-methamphetamine (a sympathomometic drug for treating sinus disorder) as *d*-methamphetamine (a widely abused stimulant) mandates the development of reliable procedures for the determination of these enantiomers. Although radioimmunoassay, nuclear magnetic resonance (NMR), and high-performance liquid chromatography (HPLC) can be used for the determination of enantiomers, MS-based methods appear to be more versatile.

The combined use of chiral and achiral column GC and MS have been used for the differentiation of enantiomers of amphetamine and related drugs. The compound is first derivatized with chiral reagents, such as (R)-α-phenylbutyric acid,[86] N-pentafluorobenzoyl-S-(−)-prolyl-*l*-imidazolide,[198] or N-trifluoroacetyl-*l*-prolyl chloride (*l*-TPC).[67,68] The *l*-TPC derivatized enantiomers are diastereoisomers[67,68] which are then analyzed by a GC/MS system equipped with SP-2100 (achiral) and Chirasil-Val (chiral) capillary columns. The total resolution of four possible isomers by the chiral column facilitates the determination of not only the *d*- and *l*-amphetamine in the sample, but also the enantiomeric purities of the commercial *l*-TPC derivatizing reagent. With the purity of the *l*-TPC determined, the enantiomeric ratio of the *d*- and *l*-amphetamine can also be determined by the achiral column. This approach is successfully applied to a street sample ("White Cross") and a simulated methamphetamine preparation.

Similar approaches have also been applied to an HPLC/MS system for the determinations of enantiomeric drugs commonly found in the horseracing industry.[199]

3. Isotope Composition Measurement

Sample decomposition, successive dilution, and contamination during the storage and distribution process may hinder the application of sample differentiation methods based solely on component identification and quantitation. A complementary approach is based on isotope composition measurements. Stable isotope ratio determinations are frequently used in studying geochemical samples, carbon dioxide fixation pathways in plants, biomedical metabolism of drugs, environmental research, food adulterations, and the differentiations of explosives, and hairs. It has also been applied by this author and others to the differentiations of marijuana,[200] caffeine,[201] and diazepam (7-chloro-1,3-dihydro-1-methyl-5-phenyl-2H-1,4-benzodiazepin-2-on).[202]

The essential features of this approach are the use of an isotope ratio mass spectrometer and the complete conversion of the sample into a gas which includes the element to be determined. For carbon isotope composition determination, samples are combusted in a vacuum line. The resulting CO_2 is isolated from the combustion product

Table 10

CALCULATED AND MEASURED $^{13}C:^{12}C$
RATIO IN $(C_3H_8N)^+$ DERIVED FROM
METHAMPHETAMINE SYNTHESIZED WITH
VARYING AMOUNT OF ^{13}C-METHYLAMINE

| | | $^{13}C:^{12}C$ in $(C_3H_8N)^+$ | | |
| | | Measured | | |
^{13}C-Methylamine used (%)	Cal'd	Ave.	Std. dev.	No.
0	0.03961	0.03873	0.00068	21
0.25	0.04199	0.04140	0.00074	21
0.50	0.04439	0.04414	0.00066	21

and purified. For D:H ratio determination, the water resulting from combustion is reduced to hydrogen. For oxygen isotope ratio measurement, the sample is heated with $HgCl_2$. Oxygen in the sample is converted to CO_2 and CO. CO is further converted to CO_2 via an electric discharge. The resulting CO_2 is combined. Mass spectrometers used for isotope ratio measurements are readily available from several manufacturers. The instrument usually consists of a double inlet/double collector design to assure best measurement precision. Gas samples are introduced to the mass spectrometer with steady flow for minimal population fluctuation in the ion source.

By measuring the hydrogen and carbon isotope ratios, several production batches of the commercial drug diazepam have been differentiated.[202] Similarly, caffeine from different geographic origins have been characterized based on their carbon, hydrogen, and oxygen isotope ratios.[201] Marijuana samples grown indoors and outdoors and from urban and rural areas appear to show different carbon isotope compositions.[200] As the number of elements measured for isotope composition is increased, the chance for successful sample differentiation is also enhanced. However, one should also consider the practicality of this approach. Carbon isotope measurement is the least complicated process and appears to have a great potential for on-line measurement and practical application.

Another version of using isotope ratio measurement as a mechanism for sample differentiation involves the "isotope tagging" process. In parallel with the mechanism of labeling medically important research drugs with radioactive and nonradioactive isotopes,[203,204] the concept is presented here, that drugs with high potential for being channeled into illicit use could be labeled with specific amounts of a nonradioactive isotope during the manufacturing process. A subsequent isotope ratio measurement could then be used to differentiate samples from different sources, and to further identify the source of the drug and/or its precursors.

To illustrate this concept, an example is made of the frequently abused methamphetamine. A study[66] from this author's group incorporates varying amounts of ^{13}C-labeled methylamine during the methamphetamine synthesis process. Synthesized products are introduced into the mass spectrometer through the gas chromatograph inlet without prior clean-up. The $^{13}C:^{12}C$ ratio in the sample is monitored by an electron impact/quadrupole mass spectrometer, with selected ion monitoring the intensity ratio of the m/z 59 and the m/z 58 ions of the $(C_3H_8N)^+$ fragment. The measured ratios are in excellent agreement with the calculated ones as shown in Table 10. A variation step of 0.25% in the amount of ^{13}C-methylamine used is sufficient for product differentiation. This variation step could be reduced by at least 50-fold with the use of an isotope ratio mass spectrometer. It may well provide a sensible mechanism for monitoring controlled substances.

Table 11
QUANTITATIVE DRUG ANALYSIS

Compound category	Data acquisition	Stationary phase	Quantitation procedure	Standard used	Ref.
Morphine	GC/MS, EI, SIM	OV-1, OV-17	Internal std.	Isotope analog	206—208
Amphetamine	GC/MS, CI, SIM	Thermon-3000	Cal. curve	Parent compound	95
and related	GC/MS, EI, SIM	OV-17	Cal. curve	Parent compound	209
amine drugs	GC/MS, EI, SIM	OV-17	Internal std.	Isotope analog	210
	MS, EI, High resolution	—	Internal std.	Isotope analog	211
	GC/MS, EI, SIM	SP-2100, Chirasil-Val	Std. addition	Parent compound	67
	GC/MS, CI, SIM	OV-275	Internal std.	Isotope analog	198
Barbiturate	GC/MS, EI, SIM	OV-101	Internal std.	Isotope analog	132
	MS, EI, SIM	—	Internal std.	Isotope analog	130, 212
Cocaine	GC/MS, EI, SIM	OV-1	Internal std.	Isotope analog	150, 151
	GC/MS, EI, Scan	OV-1	Internal std.	Isotope analog	154
PCP	GC/MS, EI, SIM	SP-2250	Cal. curve	Parent compound	160
	GC/MS, CI, SIM	OV-17	Internal std.	Isotope analog	162
	GC/MS, CI, CIM	SE-30	Std. addition	Parent compound	163

C. Quantitative Analysis

The use of MS and allied methods for quantitative analysis is now almost a daily practice.[205] Quantitative aspects of drug analysis cited in this chapter are summarized in Table 11. In most applications, deuterated analogs of analytes are used as the internal standard (IS), and mass spectrometers are operated under selected ion monitoring (SIM) mode. If the quantitation experiment does not require stringent accuracy and precision, SIM and IS procedure operated under normal conditions will provide satisfactory results. However, to achieve a high degree of accuracy and precision, several intrinsic methodological considerations and MS operating parameters have to be addressed. Isotope ratio measurements, whether used for obtaining isotopic composition information or for quantitative analysis using an isotope variant as the internal standard, require the most stringent precision and accuracy. Several considerations related to this aspect are addressed here.

1. Isotope-Labeled Analog as Internal Standard

Quantitation based on isotopically labeled internal standard follows the same principle of the isotope dilution method. Readers are referred to the original reference[213] for the derivatization of the following relationship:

$$y = R + Px - Qxy \qquad (2)$$

where y is the response ratio of the ion representing the analyte and its isotopically labeled counterpart derived from the internal standard; x is the quantity of the analyte. Thus, the general form of a calibration curve, relating ratio to quantity of analyte, is a *hyperbola*. This is mainly because the analyte itself also contains a small, but significant, fraction of the isotope-labeled analog; the isotope-labeled IS is not, in all practicality, isotopically pure. Only if the contribution of the analyte to the signal accounted for the IS is neglected will Equation 2 be reduced to a linear form.

Another important factor that needs to be considered is the EI fragmentation process of a compound containing hydrogens. The unavoidable "(M — xH)" phenomena[214,215] and the naturally abundant isotopes in the analyte make one wonder how accurately the measured ion intensity truly reflects the molecular population. Direct correlations of the intensity ratio of ions, derived from the analyte and the IS, to the concentration ratio of the parent molecules also include the assumption that no isotope effect[216] occurs in the fragmentation process.

2. Instrumental Parameters

For best accuracy and precision of SIM measurements, instrumental parameters, such as instrument resolution, ion monitoring position, measuring threshold setting, and dwell time, must also be carefully considered. It has been shown that instrument drift from the peak center and fluctuations in the intensities of the selected ion beams will affect the measurement precision, mainly attributable to the greater variation in the intensities of the weaker signals detected.[210] Improper threshold settings will also affect the weaker signals to a greater extent.[217,218] To achieve best results, the signal level of the IS should be comparable with that of the analyte.[219]

For GCMS applications, problems associated with ion residence time in the ion source must also be addressed.[154,206,220,221]

D. Derivatization

Chemical derivatizations of compounds for chromatographic and related analyses have been extensively reviewed,[222] and the main reasons for using derivatives are nicely outlined in a recent review article.[223] Readers are referred to these sources for complete information. Derivatizations used for the analysis of drug samples cited by this review are summarized in Table 12. Pertinent information concerning the alternation of mass spectrometric characteristics and GCMS analysis of enantiomers through chemical derivatizations are outlined as follows.

1. Alternation of Mass Spectrometric Characteristics

Mass spectra obtained from the properly derivatized compounds often show distinct characteristics not obtainable from parent compounds. Several advantages can result from the alternation of the mass spectral characteristics through derivatization as illustrated below.

For the example shown in Figure 11,[231] improved detection of amphetamine can be obtained through SIM measurement of the m/z 140 ion. The spectrum from the parent compound exhibits low intensities of ions at higher mass range. Considering the probability of contributions from interfering compounds, the low mass m/z 44 ion is not suitable for quantitation. On the other hand, under electron capture NCI/MS conditions,[24] greater sensitivity can be achieved through the introduction, into the compound of interest, of elements of high electron affinities. This approach is parallel to the selective detection techniques commonly used in GC applications.

Mass shifts in the spectra produced by different derivatizing agents provide extremely useful information for identifying an unknown compound. For example,[230] the number of trimethylsilyl (TMS) groups attached to the parent compound is deduced based on the mass shifts as a result of replacing N,O-bis-(trimethylsilyl)-acetamine (BSA) with d_9-BSA as the derivatizing agent. This information facilitates the identification of desoxymorphine-A, monoacetyl-desoxymorphine-A, and diacetyldesoxymorphine-A as the impurities in an illicit heroin sample. The same approach is used to characterize O^6- and O^3-acetylmorphine.[230]

Compared to parent compounds, TMS derivatives of N-substituted barbiturates are found to generate less olefin radical elimination [$(M - 41)^+$ and $(M - 55)^+$]. Instead, the formation of the $(M - 15^+)$ ion is favored, thus making it easier to recognize the molecular weight of the compound under examination.[134]

Table 12

DERIVATIZATION IN DRUG ANALYSIS

Compound category	Derivatization agent	Analytical procedure	Stationary phase	Main objective of the study	Ref.
Morphine	Heptafluorobutyric acid anhydride (HFBA)	GC/MS	OV-17	Structure elucidation	63
	Hexamethyldilazane	GC/MS	OV-17	Identification	36
	Bis(trimethylsilyl)acetamide (BSA), d9-BSA, HFBA	GC/MS	OV-1	Structure elucidation	230
Amphetamine and related amine drugs	Trifluoroacetic anhydride (TFAA), HFBA	GC/MS	OV-17	Quantitation	207
	N-Methyl-bis(trifluoroacetamide)	GC/MS	Carbowax 20M-KOH	EI fragmentation	90
	Trimethylsilylketene	GC/MS	Carbowax 20M, OV-101	EI fragmentation	88
	Acetic anhydride, BSA	GC/MS	OV-1, OV-17	EI fragmentation	86
	TFAA	GC/MS	OV-17	EI fragmentation	89
	TFAA	GC/MS	Thermon-3000	Quantitation	95
	Bis(trimethylsilyl)trifluoroacetamide (BSTFA)	GC/MS	SE-30	EI fragmentation	81, 93
	N-Trifluoroacetyl-l-prolyl-chloride	GC/MS	SP-2100, SE-54 Chirasil-Val	Enantiomeric separation	67, 68, 84
	(R)-α-Phenylbutyric acid	GC/MS	OV-1, OV-17	Enantiomeric separation	86
	N-Pentafluorobenzoyl-S-(−)-prolyl-l-imidazolide	GC/MS	OV-275	Enantiomeric separation	198
	1-Naphthoic acid	HPLC/MS	Pirkle 1A, Baker-bond DNBPG	Enantiomeric separation	199
	Dansyl chloride	MS	—	Quantitation	211
	TFAA	GC/MS	OV-17	Detection and quantitation	210
	TFAA and analogs, p-trifluoromethylbenzoyl chloride, pentafluorobenzoyl chloride, trichloroacetyl chloride	GC/MS	OV-1, SE-30	Comparison of agents	224
Cannabinoid	t-Butydimethylsilyl chloride, trimethylsilylketene	GC/MS	OV-101	Separation and quantitation	225

39

	Derivative	Column	Method	Application	Ref.
	BSTFA, trialkylchlorosilanes	SE-30	GC/MS	Structure elucidation	226
	BSTFA, d$_{18}$-BSTFA	SE-30	GC/MS	Structure elucidation	115, 226
	BSTFA	OV-17	GC/MS	Structure elucidation	227
	Dansyl chloride	—	CI	Structure elucidation	228
	Alkylboronic acid, BSTFA, trialkylchlorosilanes	SE-30	GC/MS	Structure elucidation	229
Barbiturate	Diazomethane	SE-52	GC/MS	EI fragmentation	127
	Diazomethane	—	MS	Quantitation	212
	Trimethylanilinium hydroxide	OV-1	GC/MS	Structure elucidation	128
	Methyl iodide, methylsulfinylmethide	OV-1	GC/MS	Structure elucidation	129
	BSTFA, d$_{18}$-BSTFA	XE-60, OV-17, Pentasil, Dexil 300	GC/MS	Structure elucidation	130
	Diazomethane, diazoethane	SE-30	GC/MS	EI fragmentation	131
	Diazomethane, diazoethane	OV-101	GC/MS	EI fragmentation	132
	BSA	—	MS	Structure elucidation	134
Cocaine	BSA	OV-1	GC/MS	EI fragmentation	230

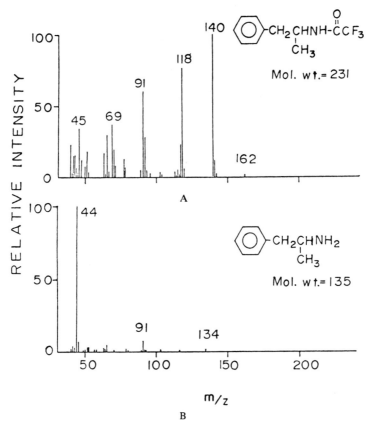

FIGURE 11.   Comparison of the EI mass spectra of (A) amphetamine and (B) ·
amphetamine trifluoroacetamide. (From Foltz, R. L., *Handbook of Mass Spectra of Drugs,* Sunshine, I., Ed., CRC Press, Boca Raton, Fla., 1981, 7. With
permission.)

2. Derivatization of Enantiomers with a Chiral Agent

MS, in general, is incapable of differentiating enantiomers. Although chiral stationary phases[232] are available for direct separation of enantiomers, derivatizations with
chiral or achiral agents are often needed for better chromatographic properties. The
derivatization of amphetamine and methamphetamine with *l*-TPC for GCMS enantiomeric analyses has been emphasized.[67,68] Total ion chromatograms of the derivatization products are shown in Figure 12. The four possible isomers resulting from the
reaction of *d*- and *l*-amphetamine with *d*- and *l*-TPC are completely resolved by the
Chirasil-Val column (Figure 12A). The methamphetamine counterparts are resolved
into three peaks (Figure 12B). Since the achiral column (SP-2100) cannot resolve enantiomers, only two peaks are observed (Figure 12C).

ACKNOWLEDGMENTS

The author is grateful to T.-G. Wu and Lisa A. Fieselman for the preparation of
figures, to Steven M. Hayes for his helpful reading and suggestions of the manuscript,
to Robert F. Borkenstein, James W. Osterberg, and Joseph D. Nicol for providing the
opportunity and the environment in which many of this author's works cited here are
conducted, and to the members, of this author's group, who have participated in these
works.

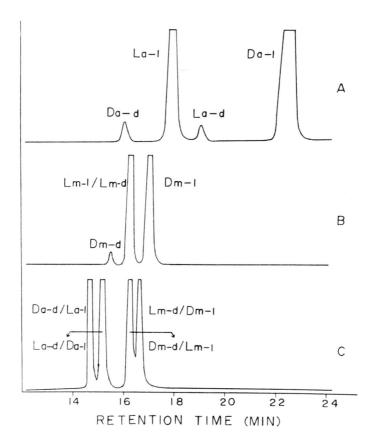

FIGURE 12. Total ion chromatograms of *l*-TPC derivatized (A) amphet-
amine, (B) methamphetamine, and (C) amphetamine and methamphetamine
mixture. (A) and (B) are obtained from a 25-m (0.30-mm I.D.) glass capillary
Chirasil-Val column; (C) is obtained from a 25-m (0.20-mm I.D.) fused silica
glass capillary SP-2100 column. La: *l*-amphetamine; Da: *d*-amphetamine;
Lm: *l*-methamphetamine; Dm: *d*-methamphetamine; *l*: *l*-TPC; *d*: *d*-TPC.
(From Liu, R. [J]. H., *Approaches to Drug Sample Differentiation,* Central
Police College Press, Taiwan, 1981, 32. With permission.)

REFERENCES

1. Zoro, J. A. and Hadley, K., Organic mass spectrometry in forensic science, *J. Forensic Sci. Soc.,* 16,
   103, 1976.
2. Yinon, J. and Zitrin, S., Processing and interpreting mass spectral data in forensic identification of
   drugs and explosives, *J. Forensic Sci.,* 22, 742, 1977.
3. Klein, M., Development of mass spectrometry as a tool in forensic drug analysis — review, in *Instrumental Applications in Forensic Drug Chemistry,* Klein, M., Kruegel, A. V., and Sobol, S. P., Eds.,
   U.S. Government Printing Office, Washington, D.C., 1978, 14.
4. Milne, G. W. A., Fales, H. M., and Law, N. C., The use of mass spectrometry for drug identification, in *Instrumental Applications in Forensic Drug Chemistry,* Klein, M., Kruegel, A. V., and Sobol,
   S. P., Eds., U.S. Government Printing Office, Washington, D.C., 1978, 91.
5. Law, N. C., Aandahl, V., Fales, H. M., and Milne, G. W. A., Identification of dangerous drugs by
   mass spectrometry, *Clin. Chim. Acta,* 32, 221, 1971.
6. Sunshine, I., Ed., *CRC Handbook of Mass Spectra of Drugs,* CRC Press, Boca Raton, Fla., 1981.
7. Mills, T., Price, W. M., Price, P. T., and Roberson, J. C., Eds., *Instrumental Data for Drug Analysis,* Vol. 1, Elsevier, Amsterdam, 1982.

8. Mills, T., Price, W. M., and Roberson, J. C., Eds., *Instrumental Data for Drug Analysis*, Vol. 2, Elsevier, Amsterdam, 1982.

9. Uelmen, G. F. and Haddox, V. G., *Drug Abuse and the Law*, 2nd ed., Clark Boardman, New York, 1983, chap. 1.

10. 21 CFR 1301.01, U.S. Government Printing Office, Washington, D.C., 1984, 84.

11. Dal Cason, T. A., Fox, R., and Frank, R. S., Investigations of clandestine drug manufacturing laboratories, *Anal. Chem.*, 52, 804A, 1980.

12. Frank, R. S., The clandestine drug laboratory situation in the United States, *J. Forensic Sci.*, 28, 18, 1983.

13. Vodra, W. W., The controlled substance act, *Drug Enforcement*, 2(2), 37, 1975.

14. Beggs, D. P. and Day, A. G., III, Chemical ionization mass spectrometry: a rapid technique for forensic analyses, *J. Forensic Sci.*, 19, 891, 1974.

15. Unger, S. E., Vincze, A., Cooks, R. G., Chrisman, R., and Rothman, L. D., Identification of quaternary alkaloids in mushroom by chromatography/secondary ion mass spectrometry, *Anal. Chem.*, 53, 976, 1981.

16. Horning, E. C., Horning, M. G., Carroll, D. I., Dzidic, I., and Stillwell, R. N., New picogram detection system based on a mass spectrometer with an external ionization source at atmospheric pressure, *Anal. Chem.*, 45, 936, 1973.

17. Foltz, R. L., Fentiman, A. F., Jr., and Foltz, R. B., *GC/MS Assays for Abused Drugs in Body Fluids*, U.S. Government Printing Office, Washington, D.C., 1980.

18. Jardine, I. and Fenselau, C., A comparison of some mass spectrometric ionization techniques using samples of morphine and illegal heroin, *J. Forensic Sci.*, 20, 373, 1975.

19. Jardine, I. and Fenselau, C., Charge exchange mass spectra of morphine and tropane alkaloids, *Anal. Chem.*, 47, 730, 1975.

20. Foltz, R. L., Quantitative analysis of abused drugs in physiological fluids by gas chromatography/chemical ionization mass spectrometry, in *Quantitative Mass Spectrometry in Life Sciences II*, De Leenheer, A. P., Foncucci, F. F., and van Peteghem, C., Eds., Elsevier, Amsterdam, 1978, 39.

21. Brandenberger, H., Negative ion mass spectrometry, in *Instrumental Applications in Forensic Drug Chemistry*, Klein, M., Kruegel, A. V., and Sobol, S. P., Eds., U.S. Government Printing Office, Washington, D.C., 1978, 48.

22. Fales, H. M., Milne, G. W. A., Winkler, H. U., Beckey, H. D., Damico, U. N., and Barron, R., Comparison of mass spectra of some biologically important compounds as obtained by various ionization techniques, *Anal. Chem.*, 47, 207, 1975.

23. Milne, G. W. A., Fales, H. M., and Axenrod, T., Identification of dangerous drugs by isobutane chemical ionization mass spectrometry, *Anal. Chem.*, 43, 1815, 1971.

24. Hunt, D. F. and Crow, F. W., Electron capture negative ion chemical ionization mass spectrometry, *Anal. Chem.*, 50, 1781, 1978.

25. Bose, A. K., Fujiwara, H., Pramanik, B. N., Lazaro, E., and Spillert, C. R., Some aspects of chemical ionization mass spectromscopy using ammonium as reagent gas: a valuable technique for biomedical and natural products studies, *Anal. Biochem.*, 89, 284, 1978.

26. Caldwell, G. and Bartmess, J. E., *In situ* generation of selective negative chemical ionization reagent gases, *Org. Mass Spectrom.*, 17, 456, 1982.

27. Madhusdanan, K. P., Use of sodium bicarbonate for *in situ* generation of reagent gas for negative chemical ionization in electron impact source, *Org, Mass Spectrom.*, 19, 517, 1984.

28. Pardanani, J. H., McLaughlin, J. L., Kondrat, R. W., and Cooks, R. G., Cactus alkaloids. XXXVI. Mescaline and related compounds from *Trichocereus peruvianus*, *Lloydia*, 40, 585, 1977.

29. Kruger, T. L., Cooks, R. G., McLaughlin, J. L., and Ranieri, R. L., Identification of alkaloids in crude extracts by mass-analyzed ion kinetic energy spectrometry, *J. Org. Chem.*, 42, 4161, 1977.

30. McClusky, G. A., Cooks, R. G., and Knevel, A. M., Direct analysis of mushroom constituents by mass spectrometry, *Tetrahedron Lett.*, 46, 4471, 1978.

31. Kondrat, R. W. and Cooks, R. G., Alkaloids in whole plant material: direct analysis by kinetic energy spectrometry, *Science*, 199, 978, 1978.

32. Youssefi, M., Cooks, R. G., and McLaughlin, J. L., Mapping of cocaine and cinnamoylcocaine in whole coca plant tissues by MIKES, *J. Am. Chem. Soc.*, 101, 3400, 1979.

33. Kondrat, R. W., McClusky, G. A., and Cooks, R. G., Multiple reaction monitoring in mass spectrometry/mass spectrometry for direct analysis of complex mixtures, *Anal. Chem.*, 50, 2017, 1978.

34. Soltero-Rigau, E., Kruger, T. L., and Cooks, R. G., Identification of barbiturates by chemical ionization and mass-analyzed ion kinetic energy spectrometry, *Anal. Chem.*, 49, 435, 1977.

35. Smith, R. M., Some applications of GC/MS in the forensic laboratory, *Am. Lab.*, 10(5), 53, 1978.

36. Kirchgessner, W. G., DiPasqua, A. C., Anderson, W. A., and Delaney, G. V., Drug identification by the application of gas chromatography/time-of-flight mass spectrometer technique, *J. Forensic Sci.*, 19, 313, 1974.

37. Sullivan, R. C., Dugan, S., Fava, J., McDonnel, E., and Yarchak, M., Evaluation and selection of gas chromatography/mass spectrometry systems for the identification of dangerous drugs, *J. Police Sci. Admin.*, 2, 185, 1974.
38. Callieux, A. and Allain, P., Superiority of chemical ionization on electron impact for identification of drugs by gas-chromatograph-mass spectrometer, *J. Anal. Toxicol.*, 3, 39, 1979.
39. Arpino, P. J., On-line liquid chromatography-mass spectrometry: the monitoring of HPLC effluents by a quadrupole mass spectrometer and a direct liquid inlet interface (DLI), in *Instrumental Applications in Forensic Drug Chemistry*, Klein, M., Kruegel, A. V., and Sobol, S. P., Eds., U.S. Government Printing Office, Washington, D.C., 1978, 151.
40. Kenyon, C. N., Melera, A., and Erni, F., Utilization of direct liquid inlet LC/MS in studies of pharmacological and toxicological importance, *J. Anal. Toxicol.*, 5, 216, 1981.
41. Henion, J. D., Thomson, B. A., and Dawson, P. H., Determination of sulfa drugs in biological fluids by liquid chromatography/mass spectrometry/mass spectrometry, *Anal. Chem.*, 54, 451, 1982.
42. Slack, J. A. and Irwin, W. J., Role of pyrolysis-gas chromatography-mass spectrometry in the analysis of drugs, *Proc. Anal. Div. Chem. Soc.*, 14, 215, 1977.
43. Hughes, J. C., Wheals, B. B., and Whitehouse, M. J., Pyrolysis mass spectrometry. A technique of forensic potential?, *Forensic Sci.*, 10, 217, 1977.
44. Down, G. J. and Gwyn, S. A., Investigation of direct thin-layer chromatography-mass spectrometry as a drug analysis technique, *J. Chromatogr.*, 103, 208, 1975.
45. Henion, J., Maylin, G. A., and Thomson, B. A., Determination of drugs in biological samples by thin-layer chromatography-tandem mass spectrometry, *J. Chromatogr.*, 271, 107, 1983.
46. Audier, H., Fetizon, M., Ginsburg, D., Mandelbaum, A., and Rull, T., Mass spectrometry of the morphine alkaloids, *Tetrahedron Lett.*, 13, 1965.
47. Nakata, H., Hirata, Y., Tatematsu, A., Toda, H., and Sawa, Y. K., Mass spectrometry of morphine alkaloids. I. Fragmentation of morphine and related compounds, *Tetrahedron Lett.*, 829, 1965.
48. Mandelbaum, A. and Ginsburg, D., Studies in mass spectrometry. IV. Steric direction of fragmentation in *cis-* and *trans-* B:C ring-fused morphine derivatives, *Tetrahedron Lett.*, 2479, 1965.
49. Wheeler, D. M. S., Kinstle, T. H., and Rinehart, K. L., Jr., Mass spectral studies of alkaloids related to morphine, *J. Am. Chem. Soc.*, 89, 4494, 1967.
50. Sastry, S. D., Alkaloids, in *Biochemical Applications of Mass Spectrometry*, Waller, G. R., Ed., John Wiley & Sons, New York, 1972, chap. 24.
51. Fales, H. M., Lloyd, H. A., and Milne, G. W. A., Chemical ionization mass spectrometry of complex molecules. II. Alkaloids, *J. Am. Chem. Soc.*, 92, 1590, 1970.
52. Saferstein, R. and Chao, J.-M., Identification of drugs by chemical ionization mass spectrometry, *J. Assoc. Anal. Chem.*, 56, 1234, 1973.
53. Saferstein, R., Chao, J.-M., and Manura, J., Identification of drugs by chemical ionization mass spectroscopy. II, *J. Forensic Sci.*, 19, 463, 1974.
54. Chao, J.-M., Saferstein, R., and Manura, J., Identification of heroin and its diluents by chemical ionization mass spectroscopy, *Anal. Chem.*, 46, 296, 1974.
55. Zitrin, S. and Yinon, J., Identification of opium by chemical ionization mass spectrometry, *Anal. Lett.*, 10, 235, 1977.
56. Saferstein, R., Manura, J., and Brettell, T. A., Chemical ionization mass spectrometry of morphine derivatives, *J. Forensic Sci.*, 24, 312, 1979.
57. Issachar, D. and Yinon, J., Reduction of keto acids and differentiation between isomers by chemical ionization mass spectrometry, *Anal. Chem.*, 52, 49, 1980.
58. Rohwedderm, W. K. and Cox, S. M., Hydrogenation during chemical ionization mass spectrometry, personal communication.
59. Liu, R. [J.] H., Low, I. A., Smith, F. P., Piotrowski, E. G., and Hsu, A.-F., Chemical ionization mass spectrometric characteristics of morphine alkaloids, *Org. Mass Spectrom.*, 20, 511, 1985.
60. Lin, Y. Y. and Smith, L. L., Active hydrogen by chemical ionization mass spectrometry, *Biomed. Mass Spectrom.*, 6, 15, 1979.
61. Smith, R. M., Forensic identification of opium by computerized gas chromatography/mass spectrometry, *J. Forensic Sci.*, 18, 327, 1973.
62. Nakamura, G. R., Noguchi, T. T., Jackson, D., and Banks, D., Forensic identification of heroin in illicit preparations using integrated gas chromatography and mass spectrometry, *Anal. Chem.*, 44, 408, 1972.
63. Moore, J. M. and Klein, M., Identification of O³-monoacetylmorphine in illicit heroin using gas chromatography-electron-capture detection and mass spectrometry, *J. Chromatogr.*, 154, 76, 1978.
64. Barron, R. P., Kruegel, A. V., Moore, J. M., and Kram, T. C., Identification of impurities in illicit methamphetamine samples, *J. Assoc. Off. Anal. Chem.*, 57, 1147, 1974.

65. Kram, T. C. and Kruegel, A. V., The identification of impurities in illicit methamphetamine exhibits by gas chromatography/mass spectrometry and nuclear magnetic resonance spectroscopy, *J. Forensic Sci.*, 22, 40, 1977.

66. Low, I. A., Piotrowski, E. G., Furner, R. L., and Liu, R. H., Carbon-13 tagging as a tracing mechanism — methamphetamine example, The 37th Annual Meeting of the American Academy of Forensic Sciences, Abstract, Las Vegas, Nev., Feb., 1985, 56.

67. Liu, R. J. H. and Ku, W. W., Determination of enantiomeric N-trifluoroacetyl-*l*-prolyl chloride amphetamine derivatives by capillary gas chromatography/mass spectrometry with chiral and achiral stationary phases, *Anal. Chem.*, 53, 2180, 1984.

68. Liu, R. J. H., Ku, W. W., Tsay, J. T., Fitzgerald, M. P., and Kim, S., Approaches to drug sample differentiation. III. A comparative study of the use of chiral and achiral capillary column gas chromatography/mass spectrometry for the determination of methamphetamine enantiomers and possible impurities, *J. Forensic Sci.*, 27, 39, 1982.

69. Lomonte, J. N., Lowry, W. T., and Stone, I. C., Contaminants in illicit amphetamine preparations, *J. Forensic Sci.*, 21, 575, 1976.

70. van der Ark, A. M., Verweij, A. M. A., and Sinnema, A., Weakly basic impurities in illicit amphetamine, *J. Forensic Sci.*, 23, 693, 1978.

71. Kram, T. C., Reidentification of a major impurity in illicit amphetamine, *J. Forensic Sci.*, 24, 596, 1979.

72. Theeuwen, A. B. E. and Verweij, A. M. A., Impurities in illicit amphetamine. VII. Identification of benzyl methyl ketone phenylisopropylimine and benzylmethyl ketone benzylimine in amphetamine, *Forensic Sci. Int.*, 15, 237, 1980.

73. Shulgin, A. T., Sargent, C., and Naranjo, C., Structure-activity relationship of one-ring psychotominetics, *Nature (London)*, 221, 537, 1969.

74. Poklis, A., Mackell, M. A., and Drake, W. K., Fatal intoxication from 3,4-methylenedioxyamphetamine, *J. Forensic Sci.*, 24, 70, 1979.

75. Clark, C. C., The identification of methoxy-N-methylamphetamines, *J. Forensic Sci.*, 29, 1056, 1984.

76. Soine, W. H., Shark, R. E., and Agee, D. T., Differentiation of 2,3-methylenedioxyamphetamine from 3,4-methylenedioxyamphetamine, *J. Forensic Sci.*, 28, 386, 1983.

77. Bailey, K., By, A. W., Legault, D., and Verner, D., Identification of the N-methylated analogs of the hallucinogenic amphetamines and some isomers, *J. Assoc. Off. Anal. Chem.*, 58, 62, 1975.

78. Cowie, J. S., Holtham, A. L., and Jones, L. V., Identification of the major impurities in the illicit manufacture of tryptamines and related compounds, *J. Forensic Sci.*, 27, 527, 1982.

79. Dal Cason, T. A., Angelos, S. A., and Raney, J. K., A clandestine approach to the synthesis of phenyl-2-propanone from phenylpropenes, *J. Forensic Sci.*, 29, 1187, 1984.

80. Soine, W. H., Thomas, M. N., Shark, R. E., Jane, S., and Agee, D. T., Differentiation of side chain positional isomers of amphetamine, *J. Forensic Sci.*, 29, 177, 1984.

81. Repke, D. B., Leslie, T., Mandell, D. M., and Kish, N. G., GLC-mass spectral analysis of psilocin and psilocybin, *J. Pharm. Sci.*, 66, 743, 1977.

82. White, P. C., Analysis of extracts from psilocybe semilanceata mushrooms by high-pressure liquid chromatography, *J. Chromatogr.*, 169, 453, 1979.

83. Casale, J. F., An aqueous-organic extract method for the isolation and identification of psilocin from hallucinogenic mushrooms, *J. Forensic Sci.*, 30, 247, 1985.

84. Liu, R. H., Ku, W. W., and Fitzgerald, M. P., Separation and characterization of amine drugs and their enantiomers by capillary column gas chromatography-mass spectrometry, *J. Assoc. Off. Anal. Chem.*, 66, 1443, 1983.

85. Frank, H., Nicholson, G. J., and Bayer, E., Gas chromatographic-mass spectrometric analysis of optically active metabolites and drugs on a novel chiral stationary phase, *J. Chromatogr.*, 146, 197, 1978.

86. Gilbert, M. T. and Brooks, C. J. W., Characterization of diastereomeric and enantiomeric ephedrines by gas chromatography combined with electron impact mass spectrometry and isobutane chemical ionization mass spectrometry, *Biomed. Mass Spectrom.*, 4, 226, 1974.

87. Bailey, K., Mass spectra of dimethoxyamphetamine hydrochlorides, *Anal. Chim. Acta*, 60, 287, 1972.

88. Coutts, R. T., Jones, G. R., Benderly, A., and Mak, A. L. C., A note on the synthesis and gas chromatographic-mass spectrometric properties of N-(trimethylsilyl)-acetate of amphetamine and analogs, *J. Chromatogr. Sci.*, 17, 350, 1979.

89. Coutts, R. T., Dawe, R., Jones, G. R., Liu, D. F., and Midha, K. K., Analysis of perfluoroacyl derivatives of ephedrine, pseudoephedrine and analogues by gas chromatography and mass spectrometry. I. Trifluoroacetyl derivatives, *J. Chromatogr.*, 190, 53, 1980.

90. Brettell, T. A., Analysis of N-mono-trifluoroacetyl derivatives of amphetamine analogues by gas chromatography and mass spectrometry, *J. Chromatogr.*, 257, 45, 1983.

91. Reisch, J., Pagnucco, R., Alfes, H., Jantos, N., and Hollman, H., Mass spectra of derivatives of phenylalkylamines, *J. Pharm. Pharmacol.*, 20, 81, 1968.
92. Bellman, S. W., Mass spectral identification of some hallucinogenic drugs, *J. Assoc. Off. Anal. Chem.*, 51, 164, 1968.
93. Narasimhachari, N., Spaide, J., and Heller, B., Gas liquid chromatographic and mass spectrometric studies on trimethylsilyl derivatives of N-methyl- and N,N-dimethyltryptaimes, *J. Chromatogr. Sci.*, 9, 502, 1971.
94. Marde, Y. and Ryhage, R., Negative-ion mass spectrometry of amphetamine congeners, *Clin. Chem.*, 24, 1720, 1978.
95. Suzuki, O., Hattori, H., and Asano, M., Detection of methamphetamine and amphetamine in a single human hair by gas chromatography/chemical ionization mass spectrometry, *J. Forensic Sci.*, 29, 611, 1984.
96. Lee, M. L., Novotny, M., and Bartle, K. D., Gas chromatography/mass spectrometric and nuclear magnetic resonance spectrometric studies of carcinogenic polynuclear aromatic hydrocarbons in tobacco and marijuana smoke condensates, *Anal. Chem.*, 48, 405, 1976.
97. Merli, F., Wiesler, D., Maskarinec, M. P., Novotny, M., Vassilaros, D. L., and Lee, M. L., Characterization of the basic fraction of marijuana smoke by capillary gas chromatography/mass spectrometry, *Anal. Chem.*, 53, 1929, 1981.
98. Thibault, R., Stall, W. J., Master, R. G., and Gravier, R. R., Swabbing for trace marihuana, *J. Forensic Sci.*, 28, 15, 1983.
99. Heerma, W., Terlouw, J. K., Laven, A., Dijkstra, G., Kuppers, F. J. E. M., Lousberg, J. J. C., and Salemink, C., Structure elucidation of pyrolytic products of cannabidiol by mass spectrometry, in *Mass Spectrometry in Biochemistry and Medicine*, Frigerio, A. and Castagnoli, N., Jr., Eds., Raven Press, New York, 1974, 219.
100. Salemink, C., Pyrolysis of cannabinoids, in *Marihuana: Chemistry, Biochemistry, and Cellular Effects*, Nahas, G. G., Paton, W. D. M., and Idanpaan-Heikkila, J. E., Eds., Springer-Verlag, New York, 1976, 31.
101. Kephalas, T. A., Kiburis, J., Michael, C. M., Miras, C. J., and Papadakis, D. P., Some aspects of cannabis smoke chemistry, in *Marihuana: Chemistry, Biochemistry, and Cellular Effects*, Nahas, G. G., Paton, W. D. M., and Idanpaan-Heikkila, J. E., Eds., Springer-Verlag, New York, 1976, 39.
102. Liu, R. J. H., Fitzgerald, M. P., and Smith, G. V., Mass spectrometric characterization of cannabinoids in raw *Cannabis sativa* L., *Anal. Chem.*, 51, 1875, 1979.
103. Liu, R. J. H. and Fitzgerald, M. P., Mass spectrometric differentiation of cannabinoid-containing samples, *J. Forensic Sci.*, 25, 815, 1980.
104. Nie, N. H., Hull, C. H., Jenkins, J. A., Steinbrenner, K., and Bent, D. H., *Statistical Package for the Social Sciences*, 2nd ed., McGraw-Hill, New York, 1975, 320.
105. Vree, T. B., Breimer, D. D., van Ginneken, A. M., Rossum, J. M., and de Witte, R. A., Identification of cannabivarins in hashish by a new method of combined gas chromatography-mass spectrometry, *Clin. Chim. Acta*, 34, 365, 1971.
106. Vree, T. B., Breimer, D. D., van Ginneken, C. A. M., and van Rossum, J. M., Identification in hashish of tetrahydrocannabinol, cannabidiol and cannabinol analogues with a methyl side-chain, *J. Pharm. Pharmacol.*, 24, 7, 1972.
107. Vree, T. B., Breimer, D. D., van Ginneken, C. A. M., and van Rossum, J. M., Identification of cannabicyclol with a pentyl or propyl side-chain by means of combined gas chromatography-mass spectrometry, *J. Chromatogr.*, 74, 124, 1972.
108. Brecht, C. A. L., Ch. Lousberg, R. J. J., Kuppers, E. J. E. M., Salemink, C. A., Vree, T. B., and van Rossum, J. M., Cannabis. VII. Identification of cannabinol methyl ether from hashish, *J. Chromatogr.*, 81, 163, 1973.
109. Vree, T. B., Mass spectrometry of cannabinoids, *J. Pharm. Sci.*, 66, 1444, 1977.
110. Turner, C. E., Hadley, K. W., Holley, J. H., Billets, S., and Mole, M. L., Constituents of *Cannabis sativa* L. VIII. Possible biological application of a new method to separate cannabidiol and cannabichromene, *J. Pharm. Sci.*, 64, 810, 1975.
111. Waller, C. W., Hadley, K. W., and Turner, C. E., Detection and identification of compounds in cannabis, in *Marihuana: Chemistry, Biochemistry, and Cellular Effects*, Nahas, G. G., Ed., Springer-Verlag, New York, 1976, chap. 2.
112. Turner, C. E., Bouwsma, O. J., Billets, S., and Elsohly, M. A., Constituents of *Cannabis sativa* L. XVIII. Electron voltage selected ion monitored study of cannabinoids, *Biomed. Mass Spectrom.*, 7, 247, 1980.
113. Liu, R. H., Ramesh, S., Liu, J. Y., and Kim, S., Qualitative and quantitative analysis of commercial polychlorinated biphenyl formulation mixtures by single ion monitoring gas-liquid chromatography/mass spectrometry and multiple regression, *Anal. Chem.*, 56, 1808, 1984.

114. Turner, C. E., Elsohly, M. A., and Boeren, E. G., Constitutes of *Cannabis sativa* L. XVII. A review of the natural constitutes, *J. Nat. Prod.*, 43, 169, 1980.

115. Harvey, D. J., Characterization of the butyl homologues of Δ-1-tetrahydrocannabinol, cannabinol and cannabidiol in samples of cannabis by combined gas chromatography and mass spectrometry, *J. Pharm. Pharmacol.*, 28, 280, 1976.

116. Friedrich-Fiechtl, J. and Spiteller, G., Neue Cannabinoide. I, *Tetrahedron*, 31, 479, 1975.

117. Elsohly, M. A., El-Feraly, F. S., and Turner, C. E., Isolation and characterization of (+)-cannabitriol and (−)-10-ethoxy-9-hydroxy-Δ⁶ᵃ[10a]-tetrahydrocannabinol: two new cannabinoids from *Cannabis sativa* L. Extract. *Lloydia*, 40, 275, 1977.

118. Budzikiewicz, H., Alpin, R. T., Lightner, D. A., Djerassi, C., Mechoulam, R., and Gaoni, Y., Massenspektroskopie und ihre anwendung auf Strukturelle und Stereochemische Probleme. LXVIII. Massenspektroskopische untersuchung der inhaltsstoffe von Haschisch, *Tetrahedron*, 21, 1881, 1965.

119. Claussen, U., Fehlhaber, H.-W., and Korte, F., Haschisch. XI. Massenspektrometrometrische bestimmung von Haschisch-inhaltsstoffen. II, *Tetrahedron*, 22, 3535, 1966.

120. Terlouw, J. K., Heerma, W., Burgers, P. C., Dijkstra, G., Boon, A., Kramer, H. F., and Salemink, C. A., The use of metastable ion characteristics for the determination of ion structures of some isomeric cannabinoids, *Tetrahedron*, 30, 4243, 1974.

121. Vree, T. B. and Nibbering, N. M. M., Phenolic proton transfer to the 1,2 double bond in the molecular ion of *trans*-1,2-tetrahydrocannabinol, *Tetrahedron*, 29, 3852, 1973.

122. Foltz, R. L., Fentiman, A. F., Jr., and Foltz, R. B., *GC/MS Assays for Abused Drugs in Body Fluids*, U.S. Government Printing Office, Washington, D.C., 1980, chap. 6.

123. Costopanagiotis, A. and Budzikiewicz, H., Massenspektroskopie in der drogenanalyse — die Massenspektren von Barbitursaurederivaten, *Monatsh. Chem.*, 96, 1800, 1965.

124. Grutzmacher, H.-F. and Arnold, W., Massenspektren von Barbitursaurederivaten, *Tetrahedron Lett.*, 13, 1365, 1966.

125. Coutts, R. T. and Lolock, R. A., Identification of medicinal barbiturates by means of mass spectrometry, *J. Pharm. Sci.*, 57, 2096, 1968.

126. Coutts, R. T., and Locock, R. A., The (M-43)⁺ ion in the mass spectra of some medicinal barbiturates, *J. Pharm. Sci.*, 58, 775, 1969.

127. Gilbert, J. N. T., Millard, B. J., and Powell, J. W., Combined gas-liquid chromatography-mass spectrometry in the study of barbiturate metabolism, *J. Pharm. Pharmacol.*, 22, 897, 1970.

128. Skinner, R. F., Gallaher, E. G., and Predmore, D. B., Rapid determination of barbiturates by gas chromatography-mass spectrometry, *Anal. Chem.*, 45, 574, 1973.

129. Thompson, R. M. and Desiderio, D. M., Permethylation of barbiturates, separation and characterization of the reaction products by gas chromatography-mass spectrometry, *Org. Mass Spectrom.*, 7, 989, 1973.

130. Falkner, F. C. and Watson, J. T., Mass spectrometry of the trimethylsilyl derivatives of medicinal barbiturates, *Org. Mass Spectrom.*, 8, 257, 1974.

131. Harvey, D. J., Nowlin, J., Hickert, P., Butler, C., Cansow, O., and Horning, M. G., Characterization of the isomeric dialkylbarbituric acids formed by the reaction of barbiturates with diazoalkanes, *Biomed. Mass Spectrom.*, 1, 340, 1974.

132. van Langenhove, A., Biller, J. E., Biemann, K., and Browne, T. R., Simultaneous determination of phenobarbital and p-hydroxyphenobarbital and their stable isotope labeled analogs by gas chromatography mass spectrometry, *Biomed. Mass Spectrom.*, 9, 201, 1982.

133. Rautio, M. and Lounasmaa, M., Mass spectral differentiation of some unsymmetrically substituted isomeric dihydrobarbiturates, *Acta Chem. Scand. B*, 31, 528, 1977.

134. Watson, J. T. and Falkner, F. C., The mass spectra of some N-substituted barbitals and their trimethylsilyl derivatives, *Org. Mass Spectrom.*, 7, 1227, 1973.

135. Klein, M., Mass spectral fragmentation of C₅-substituted barbituric acid derivatives, in *Recent Developments in Mass Spectrometry in Biochemistry and Medicine*, Vol. 1, Frigerio, A., Ed., Plenum Press, New York, 1978, 471.

136. Fales, H. M., Milne, G. W. A., and Axenrod, T., Identification of barbiturates by chemical ionization mass spectrometry, *Anal. Chem.*, 42, 1432, 1970.

137. Hunt, D. F. and Ryan, J. F., III, Argon-water mixtures as reagents for chemical ionization mass spectrometry, *Anal. Chem.*, 44, 1306, 1972.

138. Hunt, D. F., Stafford, G. C., Jr., Crow, F. W., and Russell, J. W., Pulsed positive negative ion chemical ionization mass spectrometry, *Anal. Chem.*, 48, 2098, 1978.

139. Jones, L. V. and Whitehouse, M. J., Anion mass spectrometry of barbiturates, *Biomed. Mass Spectrom.*, 8, 231, 1981.

140. State v. McNeal, 288 N.W. 2d 874 (Wis. App. 1980).

141. Lukaszewski, T. and Jeffery, W. K., Impurities and artifacts of illicit cocaine, *J. Forensic Sci.*, 25, 499, 1980.

142. Moore, J. M., Identification of *cis-* and *trans-*cinnamoylcocaine in illicit cocaine seizures, *J. Assoc. Off. Anal. Chem.*, 56, 1199, 1973.
143. Cooper, D. A. and Allen, A. C., Synthetic cocaine impurities, *J. Forensic Sci.*, 29, 1045, 1984.
144. Siegel, J. A. and Cormier, R. A., The preparation of *d*-pseudococaine from *l*-cocaine, *J. Forensic Sci.*, 25, 357, 1980.
145. Allen, A. C., Cooper, D. A., Kiser, W. O., and Cottrell, R. C., The cocaine diastereoisomers, *J. Forensic Sci.*, 26, 12, 1981.
146. Findlay, S. P., The three-dimensional structure of the cocaines. I. Cocaine and pseudococaine, *J. Am. Chem. Soc.*, 76, 2855, 1954.
147. Moore, J. M., Gas chromatographic detection of ecgonine and benzoylecgonine in cocaine, *J. Chromatogr.*, 101, 215, 1975.
148. Jindal, S. P., Lutz, T., and Vestergaard, P., Gas-liquid chromatographic-mass spectrometric determination of lidocaine in an illicit sample of cocaine, *J. Chromatogr.*, 179, 357, 1979.
149. Lewin, A. H., Parker, S. R., and Carroll, F. I., Positive identification and quantitation of isomeric cocaines by high-performance liquid chromatography, *J. Chromatogr.*, 193, 371, 1980.
150. Jindal, S. P. and Vestergaard, P., Quantitation of cocaine and its principal metabolite, benzoylecgonine, by GLC-mass spectrometry using stable isotope labeled analogs as internal standards, *J. Pharm. Sci.*, 67, 811, 1978.
151. Jindal, S. P., Lutz, T., and Vestergaard, P., Mass spectrometric determination of cocaine and its biological active metabolite, norcocaine, in human urine, *Biomed. Mass Spectrom.*, 5, 658, 1978.
152. Chinn, D. M., Crouch, D. J., Peat, M. A., Finkle, B. S., and Jennison, T. A., Gas chromatography-chemical ionization mass spectrometry of cocaine and its metabolites in biological fluids, *J. Anal. Toxicol.*, 4, 37, 1980.
153. Foltz, R. L., Fentiman, A. F., Jr., and Foltz, R. B., *GC/MS Assays for Abused Drugs in Body Fluids*, U.S. Government Printing Office, Washington, D.C., 1980, chap. 9.
154. Clark, C. C., Mass spectral quantitation of cocaine HCl in powders, *J. Assoc. Off. Anal. Chem.*, 64, 884, 1981.
155. Bailey, K., Gagne, D. R., and Pike, R. K., Identification of some analogs of the hallucinogen phencyclidine, *J. Assoc. Off. Anal. Chem.*, 59, 81, 1976.
156. Cone, E. J., Darwin, W. D., Yousefnejad, D., and Buchwald, W. F., Separation and identification of phencyclidine precursors, metabolites and analogs by gas and thin-layer chromatography and chemical ionization mass spectrometry, *J. Chromatogr.*, 117, 149, 1979.
157. Smialek, J. E., Monforte, J. R., Gault, R., and Spitz, U., Cyclohexamine ("rocket fuel")-phecyclidine's potent analog, *J. Anal. Toxicol.*, 3, 209, 1979.
158. Legault, D., Investigation of the gas-liquid chromatographic separation of phencyclidine and some heterocyclic analogues by combined gas-liquid chromatography-mass spectrometry, *J. Chromatogr.*, 202, 309, 1980.
159. Soine, W. H., Balster, R. L., Berglund, K. E., Martin, C. D., and Agee, D. T., Identification of a new phencyclidine analog, 1-(l-phenylcyclohexyl)-4-methylpiperidine, as a drug of abuse, *J. Anal. Toxicol.*, 6, 41, 1982.
160. Kelly, R. C. and Christmore, D. S., Identification of phencyclidine and its analogues at low concentrations in urine by selected ion monitoring, *J. Forensic Sci.*, 27, 827, 1982.
161. Wong, L. K. and Biemann, K., Metabolites of phencyclidine, *Clin. Toxicol.*, 9, 583, 1976.
162. Lin, D. C. K., Fentiman, A. F., Jr., Foltz, R. L., Forney, R. D., Jr., and Sunshine, I., Quantification of phencyclidine in body fluids by gas chromatography chemical ionization mass spectrometry and identification of two metabolites, *Biomed. Mass Spectrom.*, 2, 206, 1975.
163. Cone, E. J., Buchwald, W., and Yousefnejad, D., Simultaneous determination of phencyclidine and monohydroxylated metabolites in urine of man by gas chromatography-mass fragmentography with methane chemical ionization, *J. Chromatogr.*, 223, 331, 1981.
164. Foltz, R. L., Fentiman, A. F., and Foltz, R. B., *GC/MS Assays for Abused Drugs in Body Fluids*, U.S. Government Printing Office, Washington, D.C., 1980, chap. 3.
165. Inoue, A., Nakahara, Y., and Niwaguchi, T., Studies on lysergic acid diethylamide and related compounds. II. Mass spectra of lysergic derivatives, *Chem. Pharm. Bull.*, 20, 409, 1972.
166. Bailey, K., Verner, D., and Legault, D., Distinction of some dialkyl amides of lysergic and *iso*-lysergic acids from LSD, *J. Assoc. Off. Anal. Chem.*, 56, 88, 1973.
167. Schmidt, J., Kraft, R., and Voigt, D., Mass spectroscopy of natural products. III. Mass spectrometric comparison of lysergic acid and 9,10-dihydrolysergic acid, *Biomed. Mass Spectrom.*, 5, 674, 1978.
168. Stoll, A. and Hofmann, A., Partialsynthese von Alkaloiden vom typus des Ergobasins, *Helv. Chim. Acta*, 26, 944, 1943.
169. Barber, M., Weisbach, J. A., Douglas, B., and Dudek, G. O., High resolution mass spectrometry of the clavine and ergot-peptide alkaloids, *Chem. Ind.*, 1072, 1965.

170. Crawford, K. W., The identification of lysergic acid amide in Baby Hawaiian Woodrose by mass spectrometry, *J. Forensic Sci.,* 15, 588, 1970.

171. Nakahara, Y. and Niwaguchi, T., Studies on lysergic acid diethylamide and related compounds. I. Synthesis of *d*-N⁶-demethyl-lysergic acid diethylamide, *Chem. Pharm. Bull.,* 19, 2337, 1971.

172. Vokoun, J. and Rehacek, Z., Mass spectra of ergot peptide alkaloids, *Collect. Czech. Chem. Commun.,* 40, 1731, 1975.

173. Urich, R. W., Bowerman, D. L., Wittenberg, P. H., McGaha, B. L., Schisler, D. K., Anderson, J. A., Levisky, J. A., and Pflug, J. L., Mass spectral studies of ultraviolet irradiated and nonirradiated lysergic acid diethylamide extracts from illicit preparations, *Anal. Chem.,* 47, 581, 1975.

174. Stinson, S., Structure of bogus "China White" solved, *Chem. Eng. News,* 59(3), 71, 1981.

175. Kram, T. C., Cooper, D. A., and Allen, A. C., Behind the identification of China White, in *The Analytical Approach,* Grasselli, J. G., Ed., American Chemical Society, Washington, D.C., 1983.

176. Van Bever, Q. F. M., Niemegeers, C. J. E., and Janssen, P. A. J., Synthetic analgesics. Synthesis and pharmacology of the diastereoisomers of N-[3-methyl-1-(2-phenylethyl)-4-piperidyl]-N-phenylpropanamide and N-[3-methyl-1-(1-methyl-2-phenylethyl)-4-piperidyl]-N-phenylpropanamide, *J. Med. Chem.,* 17, 1047, 1974.

177. Gardocki, J. F. and Yelnosky, J., A study of some of the pharmacologic actions of fentanyl citrate, *Toxicol. Appl. Pharmacol.,* 6, 48, 1964.

178. Frincke, J. M. and Henderson, G. L., The major metabolite of fentanyl in the horse, *Drug Metab. Dispos.,* 8, 425, 1980.

179. Cheng, M. T., Kruppa, G. H., McLafferty, F. W., and Cooper, D. A., Structural information from tandem mass spectrometry for China White and related fentanyl derivatives, *Anal. Chem.,* 54, 2204, 1982.

180. Dal Cason, T. A., Angelos, S. A., and Washington, O., Identification of some chemical analogues and positional isomers of methaqualone, *J. Forensic Sci.,* 26, 793, 1981.

181. Bonnichsen, R., Fri, C.-G., Negoita, C., and Ryhage, R., Identification of methaqualone metabolites from urine extract by gas chromatography-mass spectrometry, *Clin. Chim. Acta,* 40, 309, 1972.

182. Ghaderl, S., Kulkarnl, P. S., Ledford, E. B., Jr., Wilkins, C. L., and Gross, M. L., Chemical ionization in Fourier transform mass spectrometry, *Anal. Chem.,* 53, 428, 1981.

183. Yost, R. A., Perchalski, R. J., Brotherton, H. O., Johnson, J. V., and Budd, M. B., Pharmaceutical and clinical analysis by tandem mass spectrometry, *Talanta,* 31, 929, 1984.

184. Brotherton, H. O. and Yost, R. A., Rapid screening and confirmation for drugs and metabolites in racing animals by tandem mass spectrometry, *Am. J. Vet. Res.,* 45, 2436, 1984.

185. Perchalski, R. J., Yost, R. A., and Wilder, B. J., Structural elucidation of drug metabolites by triple-quadrupole mass spectrometry, *Anal. Chem.,* 54, 1466, 1982.

186. Brotherton, H. O. and Yost, R. A., Determination of drugs in blood serum by mass spectrometry/mass spectrometry, *Anal. Chem.,* 55, 549, 1983.

187. Nicol, J. D., The bachelor of science in criminalistics, Department of Criminal Justice, University of Illinois, Chicago, 1972, 2.

188. Liu, R. J. H., Approaches to drug sample differentiation. I. A conceptual review, *J. Forensic Sci.,* 26, 651, 1981.

189. Lomonte, J. N., Lowry, W. T., and Stone, I. C., Contaminations in illicit amphetamine preparations, *J. Forensic Sci.,* 21, 575, 1976.

190. Hufsey, J. and Cooper, D., Synthetic cocaines — an overview of synthesis and analysis, *Microgram,* 12, 231, 1979.

191. van der Slooten, E. P. J. and van der Helm, H. T., Analysis of heroin in relation to illicit drug traffic, *Forensic Sci.,* 6, 83, 1975.

192. Narayanaswami, G. H. C. and Dua, R. D., Assay of major and minor constituents of opium samples and studies of their origin, *Forensic Sci. Int.,* 14, 181, 1979.

193. Huizer, H., Analytical studies on illicit heroin. II. Comparison of samples, *J. Forensic Sci.,* 28, 40, 1983.

194. O'Neil, P. J., Baker, P. B., and Gough, T. A., Illicitly imported heroin products: some physical and chemical features indicative of their origin, *J. Forensic Sci.,* 29, 889, 1984.

195. Novotny, M., Lee, M. L., Low, C.-E., and Raymond, A., Analysis of marihuana samples from different origins by high-resolution gas-liquid chromatography for forensic application, *Anal. Chem.,* 48, 24, 1976.

196. Furner, R. L., Brown, G. B., and Scott, J. W., A method for differentiation and analysis of quinine and quinidine by gas chromatography/mass spectrometry, *J. Anal. Toxicol.,* 5, 275, 1981.

197. Solomon, M. D. and Wright, J. A., False-positive for (+)-methamphetamine, *Clin. Chem.,* 23, 1504, 1977.

198. Matin, S. B., Wan, S. H., and Knight, J. B., Quantitative determination of enantiomeric compounds. I. Simultaneous measurement of the optical isomers of amphetamine in human plasma and saliva using chemical ionization mass spectrometry, *Biomed. Mass Spectrom.*, 4, 118, 1977.

199. Crowther, J. B., Covey, T. R., and Henion, J. D., Liquid chromatographic/mass spectrometric determination of optically active drugs, *Anal. Chem.*, 56, 2921, 1984.

200. Liu, R. J. H., Lin, W. F., Fitzgerald, M. P., Sexana, S. C., and Shieh, Y. N., Possible characterization of samples of *Cannabis sativa* L. by their carbon isotopic distribution, *J. Forensic Sci.*, 24, 814, 1979.

201. Dunbar, J. and Wilson, A. T., Determination of geographic origin of caffeine by stable isotope analysis, *Anal. Chem.*, 54, 590, 1982.

202. Bommer, P., Moser, H., Stichler, W., Trimborn, P., and Vetter, W., Determination of the origin of drugs by measuring natural isotope contents: D/H and $^{13}C/^{12}C$ ratio of some diazepam samples, *Z. Naturforsch.*, 31c, 111, 1976.

203. Haskins, N. J., The application of stable isotope in biomedical research, *Biomed. Mass Spectrom.*, 9, 269, 1982.

204. Baillie, T. A., The use of stable isotopes in pharmacological research, *Pharmacol. Rev.*, 33, 81, 1981.

205. Garland, W. A. and Powell, M. L., Quantitative selective ion monitoring (QSIM) of drugs and/or drug metabolites in biological matrices, *J. Chromatogr. Sci.*, 19, 392, 1981.

206. Jerpe, J. H., Bena, F. E., and Morris, W., GC-Quadrupole mass fragmentography of heroin, *J. Forensic Sci.*, 20, 557, 1975.

207. Ebbighausen, W. O. R., Mowat, J. H., Vestergaard, P., and Kline, N. S., Stable isotope method for the assay of codeine and morphine by gas chromatography-mass spectrometry. A feasibility study, in *Advances Biochem. Psychopharm.*, Vol. 7, Mendlewicz, J., Coopen, A., and van Praag, H. M., Eds., Raven Press, New York, 1973, 135.

208. Jindal, S. P., Lutz, T., and Vestergaard, P., Gas chromatographic-mass spectrometric determination of etorphine with stable isotope labeled internal standard, *Anal. Chem.*, 51, 269, 1979.

209. Frigerio, A., Fanelli, R., and Danieli, B., Detection of 2,5-dimethoxy-4-methylamphetamine by gas chromatography-mass fragmentography, *Chem. Ind.*, 769, 1972.

210. Cho, A. K., Lindeke, B., Hodshon, B. J., and Jenden, D. J., Deuterium substituted amphetamine as an internal standard in a gas chromatographic/mass spectrometric (GC/MS) assay for amphetamine, *Anal. Chem.*, 45, 570, 1973.

211. Danielson, T. J. and Boulton, A. A., Detection and quantitative analysis of amphetamine, *Biomed. Mass Spectrom.*, 1, 159, 1974.

212. Lee, M. G. and Millard, B. J., A comparison of unlabelled and labelled internal standards for quantification by single and multiple ion monitoring, *Biomed. Mass Spectrom.*, 2, 78, 1975.

213. Thorne, G. C., Gaskell, S. J., and Payne, P. A., Approaches to the improvement of quantitative precision in selected ion monitoring: high resolution applications, *Biomed. Mass Spectrom.*, 11, 415, 1984.

214. Low, I. A., Liu, R. H., Fish, F., Barker, S. A., Settine, R. L., Piotrowski, E. G., Damert, W. C., and Liu, J. Y., Selected ion monitoring mass spectrometry: parameters affecting quantitative determination, *Biomed. Mass Spectrom.*, 12, 633, 1986.

215. Benz, W., Accuracy of isotopic label calculations for spectra with a (molecular ion — hydrogen) peak, *Anal. Chem.*, 52, 248, 1980.

216. Derrick, P. J., Isotope effects in fragmentation, *Mass Spectrom. Rev.*, 2, 285, 1983.

217. Millard, P. J., *Quantitative Mass Spectrometry*, Heyden, London, 1978, 51.

218. Liu, R. H., Smith, F. P., Low, I. A., Piotrowski, E. G., Damert, W. C., Phillips, J. G., and Liu, J. Y., Direct mass spectrometric determination of carbon-13 enrichment of organic compounds, *Biomed. Mass Spec.*, 12, 638, 1986.

219. Farjo, K. and Haase, G., "Stable isotope mixing", a proposal to analyze stable isotope using a technique similar to the isotope dilution analysis. II. Mass spectrometrical H/D and N-isotope analysis, *Isotopenpraxis*, 16, 137, 1980.

220. Furner, R. L., Low, I. A., Piotrowski, E. G., and Liu, R. H., GC/Selected ion MS monitoring of isotope ratio in ^{13}C-labeled methamphetamine, 189th American Chemical Society National Meeting, Miami Beach, Fla., May, 1985.

221. Mathews, D. E. and Hayes, J. M., Systematic errors in gas chromatography-mass spectrometry isotope ratio measurements, *Anal. Chem.*, 48, 1375, 1976.

222. Knapp, D. R., *Handbook of Analytical Derivatization Reactions*, John Wiley & Sons, New York, 1979.

223. Brooks, J. W., Edmonds, C. G., Gaskell, S. J., and Smith, A. G., Derivatives suitable for GC-MS, *Chem. Phys. Lip.*, 21, 403, 1978.

224. Anggard, E. and Hankey, A., Derivatives of sympathomimetic amines for gas chromatography with electron capture detection and mass spectrometry, *Acta Chem. Scand.*, 23, 3110, 1969.

225. Knaus, E. E., Coutts, R. T., and Kazakoff, C. W., The separation, identification, and quantitation of cannabinoids and their *t*-butyldimethylsilyl, trimethylsilylacetate, and diethylphosphate derivatives using high-pressure chromatography, and mass spectrometry, *J. Chromatogr. Sci.,* 14, 525, 1976.

226. Harvey, D. J. and Paton, W. D. M., Use of trimethylsilyl and other homologous trialkylsilyl derivatives for the separation and characterization of mono- and dihydroxycannabinoids by combined gas chromatography and mass spectrometry, *J. Chromatogr.,* 109, 73, 1975.

227. Billets, S., El-Feraly, F., Fetterman, P. S., and Turner, C. E., Constituents of *Cannabis sativa L.* XII. Mass spectral fragmentation patterns for some cannabinoid acids as their TMS derivatives, *Org. Mass Spectrom.,* 11, 741, 1876.

228. Maseda, C., Yuko, M. P., Kimura, K., and Matsubara, K., Chromorphoric labeling of cannabinoids with 4-dimethylaminoazobenzene-4'-sulfonyl chloride, *J. Forensic Sci.,* 28, 911, 1983.

229. Harvey, D. J., Cyclic alkylboronates as derivatives for the characterization of cannabinolic acids by combined gas chromatography and mass spectrometry, *Biomed. Mass Spectrom.,* 4, 88, 1977.

230. Moore, J. M., The application of derivatization techniques in forensic drug analysis, in *Instrumental Applications in Forensic Drug Chemistry,* Klein, M., Kruegel, A. V., and Sobol, S. P., Eds., U.S. Government Printing Office, Washington, D.C., 1978, 180.

231. Foltz, R. L., Mass spectrometry, in *Handbook of Mass Spectra of Drugs,* Sunshine, I., Ed., CRC Press, Boca Raton, Fla., 1981, 7.

232. Liu, R. H. and Ku, W. W., Chiral stationary phases for the gas-liquid chromatographic separation of enantiomers, *J. Chromatogr.,* 271, 309, 1983.

Chapter 2

MASS SPECTROMETRY OF DRUGS AND TOXIC SUBSTANCES IN BODY FLUIDS

Michael Klein

TABLE OF CONTENTS

I. INTRODUCTION

Mass spectrometry (MS), a complex discipline used in diverse theoretical and applied fields of study, can provide the forensic scientist with evidence of drug abuse or overdose, poisoning, or environmental toxicity. The following discussion focuses on forensic applications involving mass spectrometric analysis of drugs and toxic substances in body fluids.

The scientific investigator in forensic laboratories, toxicology laboratories, and hospital emergency rooms is frequently required to analyze biological specimens for a drug or poison and/or their metabolites. Making determinations of drugs from biological fluids can be a formidable task. Often, interferences can mask the presence of a drug and its metabolites. Extracts of biological systems include numerous endogenous components and necessitate extensive procedures to isolate, or at least to clarify, individual drugs prior to identification. As a consequence, both quantitative and qualitative analysis may suffer due to drug losses. Additionally, some groups of drugs are so potent that therapeutic effectiveness at even very low plasma concentrations can easily be attained. As such, many forensic studies for drugs or other substances require sensitive and specific determination procedures.

Methods such as radioimmunoassay (RIA) or enzyme immunoassay (EIA) offer very high sensitivity; however, because of cross-reactivity, only the drug class and not the specific agent involved may be identified. It is therefore often difficult for the analyst to unequivocally distinguish a drug from its metabolites or other structurally similar substances. Other analytical methods, such as gas chromatography (GC) or high-pressure liquid chromatography (HPLC), which utilize nonspecific detectors, cannot unequivocally confirm the identification of a particular drug. In contrast, MS provides excellent specificity and sensitivity for qualitative and quantitative drug analysis.

II. ANALYSIS BY MASS SPECTROMETRY

Development of improved analytical techniques in the determination of potential or established "drugs of abuse" requires the fast, accurate and precise, highly sensitive and specific, qualitative, and quantitative determination of increasingly lower levels of drugs, drug impurities, and drug metabolites. The compound being analyzed may be in the medium of the drug sample itself, or in extracts of biological fluids (blood, urine, sweat, saliva, milk, cerebrospinal, and synovial) from individuals who have received the drug. Optimum utilization of the spectrometric analysis may be dependent upon interactive computer systems, chromatographic analysis, and chemical techniques, e.g., derivatization and extraction procedures.

By introduction of compounds eluting from a gas chromatographic column into the mass spectrometer, spectral data collected on each peak makes possible their positive identifications. GC is the most common method routinely available for resolving into individual components the highly complex mixtures of compounds encountered in drug or biological specimens.[1] The liquid chromatograph/mass spectrometer interface has advantages over GC/MS in the analysis of thermally unstable molecules.[2,3] Development of sophisticated instrumentation for GC and MS has made it possible to use these tools separately or in conjunction with a computer (COM). The methodology can be improved by the incorporation of digital computers into the system to allow automatic collection and analysis of mass spectra.[3]

The detection by GC/MS/COM of specific compounds has been developed to the point such that the system, often relied upon as a routine method in the laboratory, offers advantages of speed, reliability, and dependability for the analysis of volatile and nonvolatile materials. Most common instrumental methods do not provide the necessary levels of sensitivity, as well as specificity, for the identification and quantification of drug components. Fluorometry, RIAs, thin-layer chromatography (TLC), and many spectrophotometric methods often may not provide the adequate sensitivity and/or necessary specificity. The hazards and ethical responsibilities associated with measurement of radiolabeled drugs limit this application. Colorimetric procedures provide adequate sensitivity, but lack the specificity of mass spectrometric analysis.[4]

Even gas chromatographic methods with the flame ionization detector (FID) seldom meet the sensitivity requirements. The maximum usable sensitivity of GC (FID) would be around 0.05 μg with biological samples. Under these conditions, the flame tends to be somewhat noisy, however, and interfering peaks limit sensitivity. In addition, peaks of interest may be masked (or only partially resolved) by biological components present in the sample.[5]

Nitrogen detectors or electron capture detectors (ECD) meet sensitivity requirements, but adequate chromatographic separation of structurally similar drugs is difficult. Another disadvantage of ECD/GC is that no structural information would be associated with a response, and slight differences in conditions may effect that response. These disadvantages may lead to mistakes both in identity and quantification of material.

Even mass fragmentographic methods cause problems when ions of sufficient intensity but of low specifity are measured, i.e., when the most intense fragment ions are of low mass or are formed from structurally similar molecules. However, the advantage of using GC/MS assays with chemical ionization (CI) is that the most characteristic ion of the compound, i.e., the parent ion, would be available for quantification.[6]

MS has been shown overall to be sensitive and specific enough for monitoring concentrations of various drugs in whole blood or other photometric determinations; quite tedious extraction procedures are often necessary to minimize these effects. By using mass fragmentography, losses in extraction or chromatographic absorption can be allowed for by use of the appropriate deuterium-labeled standard.[8]

A. Quantitative Methods — Mass Fragmentography

Combined GC/MS has become an established method for detecting and quantifying very low levels of compounds in complex mixtures. The technique of mass fragmentography is one of the most sensitive detection systems known. Mass fragmentography, also known as single or multiple ion mass detection, is the simultaneous monitoring of one or more fragment ions rather than the scanning of the total ion spectrum as in conventional MS.[9,10] The use of mass fragmentography as a single or multiple ion monitoring for GC has been applied in the qualitative and quantitative analysis of low levels of endogenous compounds, drugs, drug metabolites, biological samples, pesticides, and environmental pollutants. The ion-detecting device of the mass spectrometer is set to monitor, as a function of time, a fragment ion of relatively high abundance in the spectrum of the compound of interest.

The resulting record is that of a gas chromatogram performed with a selective detector. Depending on the ions chosen, a limited number of peaks will be observed at the retention time of each compound being analyzed. Therefore, the chromatogram is relatively free of interferences from other compounds in the sample, and the method often offers sensitivity greater than that obtainable with conventional GC detectors.

Due to monitoring of specific mass ions of the compound being analyzed, a preceding separation of other drugs is not required. Quantitative GC requires that the compound being analyzed and the internal standard be completely resolved, but this is not needed in mass fragmentography.[11]

By focusing on a fixed m/z ratio (selected ion monitoring [SIM]), the mass spectrometer behaves as a highly specific gas chromatographic detector, only responding to compounds that, on fragmentation, yield ions at the m/z ratio upon which it is focused.

1. Single Ion Monitoring

Single ion monitoring usually requires selection of an appropriate derivative of a component which has at least one unique ion of high abundance at a reasonably high m/z ratio so that interference is minimized. Calibration curves must be obtained to determine the losses on extraction, purification, and derivatization, as well as losses from column absorption and variation in instrumental response. This is often time-consuming and unreliable.

The sensitivity of detection over that of conventional GC methods, e.g., electron capture, flame detection, and thermal conductivity, is 10^3 to 10^4.[10] High resolution can be used when more than one fragment ion with the same nominal m/z ratio is present. When a number of compounds in a sample and an internal standard produce on ionization an ion fragment with the identical m/z ratio, but have different GC retention times, quantitation of more than one compound is performed. The mass spectrometer is acting as a specific detector of certain structural features in a mixture. This has been referred to as "functional group analysis".

2. Multiple Ion Monitoring

The mass spectrometer focuses on several unique prominent fragment ions. Multiple ion monitoring is useful for the simultaneous quantitation of drugs and their metabolites. Maximum sensitivity results from monitoring a single ion. As more ions are selected, each will be monitored for a shorter period with subsequent decrease in signal-to-noise ratio and increased contribution from random electrical noise. However, multiple ion monitoring has the added advantage of greater specificity by focusing on more than one characteristic fragment. One application can be for detection of drugs and metabolites containing elements with more than one naturally occurring stable isotope, e.g., Cl and Br.

In carrying out quantitative estimations, the ideal standard should be as chemically similar to, but in some way distinguishable from, the compound under investigation. ^2H-, ^{13}C-, and ^{15}N-labeled compounds are the most suitable, and should be added prior to extraction and analysis. Generally, labeling with ^{13}C would be preferred over deuterium since the latter may easily exchange with hydrogen; in addition, the isotope effect of ^{13}C is extremely small. However, the natural abundance of ^{13}C is much greater than that of ^2H and this would decrease the sensitivity of the method.[12] In addition, the deuterated compounds are less expensive.

Since the chemical properties of labeled drugs are very similar to those of the parent compound, this procedure overcomes problems caused by incomplete extraction and derivatization and also minimizes adsorption losses encountered on GC.[13] In addition, they show similar fragmentation patterns to the unlabeled compound. Due to its extra mass, the resultant fragments will have different m/z ratios, distinguishing them from the unlabeled fragments. Suitably labeled compounds are therefore useful for quantifying the compound. One disadvantage is that multiple labeled standards can show GC or HPLC properties slightly different from those of the compound of interest.

When compounds other than stable isotope-labeled analogs are used as internal standards, it is necessary to determine a response factor, which is the ratio of the peak area per microgram of drug divided by the peak area per microgram of internal standard. The response factors should be constant over the range of concentrations measured. In the case of unavailable standards, a hydrocarbon or drug with suitable GC properties can be added as an internal standard.[14]

B. Common Ionization Techniques

Often, the success of a mass spectrometric analysis depends upon the type of ion source chosen. Electron impact mass spectrometry (EI/MS) of the compounds does not always provide the necessary level of confidence needed in making molecular assignments. Therefore, alternative ionization techniques can be used to further support the results.

The alternative methods of ionization most often provide simpler mass spectra dominated by a few highly intense ions. The most frequently used is chemical ionization MS (CI/MS). Neither technique is necessarily superior to the others, but the method of analysis depends on the type of compound involved and the specific problem being investigated.

1. Electron Impact Ionization

Several derivatization techniques have been studied with respect to their effect on GC and resultant EI mass spectral characteristics. These studies have been critical with respect to analysis of many drug classes. Since the major function of derivatization is to enhance volatility and thermal stability, efforts have been directed to the development of new derivatives which also improve the compatibility of various chromatographic and mass spectral techniques. With respect to improvement of MS techniques, the goal is for greater structural information from MS and improved detectability with GC detectors or with GC/MS SIM.

2. Chemical Ionization

CI/MS provides a larger fraction of ions related to the molecular weight of the sample under analysis. Other advantages are that the CI source can take most or all of the column effluent, and the carrier gas can be used as the reagent gas. CI/MS is best utilized in conjunction with EI/MS in order to complement molecular weight information with mass spectral structural information.[15] The occurrence of both proton transfer and addition reactions on the same compound can often provide further evidence for the molecular weight assigned. This results in the quasimolecular ion, $(MH)^+$, an even electron species which, if unstable, can further fragment to lose a neutral molecule. One would expect more fragmentation as the proton affinity of the reagent gas decreases. Proton abstraction $(M - H)^+$ can occur mostly with compounds of low proton affinity, e.g., alkanes. Addition reactions between functional groups and the reagent gas can also be observed.

SIM analysis using CI for various compounds has been reported to achieve comparable sensitivity with a specificity not equalled by other analytical methods. SIM analyses allow quantitation of nanogram or picogram amounts of several drugs and drug metabolites in a single run. CI is now frequently used for the quantitation of drugs and metabolites. Compounds which cannot be derivatized easily or which show poor chromatographic properties can be quantitated using the CI/MS direct insertion-SIM method.

The degree of interference depends on the uniqueness of the ion masses monitored, and, therefore, CI offers significant advantages over the more commonly used EI ionization due to the high abundance of quasimolecular ions. The ion current from the compound being analyzed is concentrated in relatively few high m/z values where background contributions are low. CI has the additional advantage of eliminating the need for a separator between the gas chromatograph and the mass spectrometer. The separator can be a major source of sample loss, owing to adsorption, decomposition, or diffusion. CI/MS offers the advantage of determining isotope ratios on compounds obtained from biological sources by improving the uniqueness of a particular ion and thus minimizing interferences from ions due to natural biological contaminants.[13] Several disadvantages of CI/MS are possible; these include sensitivity to temperature changes, which affect the extent of fragmentation, and, as the ion source gets contaminated, gradual deterioration of linearity and precision can be expected.[6]

a. Reagent Gases

A reagent gas can be used to identify basic groups in an unknown molecule. In addition, if the functional groups present in a molecule are known, a reagent gas can be selected which will selectively protonate one of these, and, therefore, provide fragmentations which are initiated by and are specific for that particular functional group. Hydrogen, methane, and isobutane all enhance the MH^+ ion and can be used in GC analysis. Water has been shown to be useful in direct determination of organic compounds in aqueous solution, producing typical ions $(MH)^-$ and $(M + H_3O)^+$. Deuterium oxide (D_2O) can additionally be used in the determination of the number of active hydrogens in a compound.

Ammonia (NH_3) and alkylamines ($R-NH_2$) selectively protonate amines, amides, and α,β-unsaturated ketones and form addition products $(M + NHR)^+$ from ketones, aldehydes, acids, and esters. When ammonia was used as the reagent gas, CI/MS of triglycerides produced peaks which corresponded to quasimolecular ions. When isobutane was the reagent gas, quasimolecular ions were usually not recorded. With ammonia as the reagent gas, the $(M + NH_4)^+$ ion was the base peak.[16] An alternative to CI/MS with ammonia is isobutane modified with ethanolamine and ethylenediamine, in which less fragmentation is obtained than with isobutane used alone.

Another method of selective detection by CI/MS is the reactant ion monitoring technique.[17] As each component in a sample elutes from the GC into the source of the mass spectrometer, there is a decrease in the ion current of the reactant ions because of the reactions of these ions with the added sample. The GC trace produced is called "reactant ion monitoring" and it is equivalent to a plot of total sample ion current vs. time.

b. Charge Exchange Gases

Charge exchange (CE) reactions arise when the ionized reagent gas cannot donate a proton. The ionization potential of the molecule must be less than the recombination energy of the ionized gas. The degree of fragmentation will depend on the energy difference between the two parameters. Combined CE/CI occurs when a second gas containing hydrogen is used with the CE gas. The second gas, once ionized, can then ionize the sample molecule. The spectra of molecules ionized in this way will present a situation between that obtained for true EI and CI spectra.[15]

Generally, CE gases (He, Ar, N_2) yield spectra which do not differ significantly from those obtained by EI. Mass spectra obtained with nitric oxide, however, show relatively abundant $(M + NO)^+$ ions with molecules containing π-electrons. Therefore, alkenes can be distinguished from alkanes and cycloalkanes since only the former give $(M + NO)^+$ ions.

Nitric oxide CI can also differentiate between primary, secondary, and tertiary alcohols. Primary alcohols give spectra in which $(M - 2)$, $(M - 3)$, and $(M - 2 + NO)$ ions are present. Secondary alcohols give major ions $(M - 1)$, $(M - 17)$, and $(M - 2 + NO)$, whereas $(M - 17)$ ions only are produced from tertiary alcohols. The mass spectra of 12 morphine and tropane alkaloids obtained by using N_2/NO as the reagent gas all showed the M^+ as the base peak.[18] In general, acetoxy, hydroxy, carbonyl, and aromatic groups will not be involved in CE since their ionization potentials are greater than the recombination energy of the NO^+ ion. Therefore, fragmentations characteristic of these groups will not occur to the same extent as in the EI spectra.

EI was shown to be 4 times more sensitive than CI and 20 times more sensitive than CE by measuring the relative intensity of the MH^+ molecular ion of heroin by SIM. The use of mixtures of CI and CE gases also offers some advantages. Mixtures of Ar/H_2 and Ar/CH_4 have been shown to distinguish mass spectrometrically between the two isomeric compounds, amobarbital and pentobarbital. The major CE gases are Ar (combined with NO, H_2O, or CH_4), He (combined with H_2O or NO), N_2 (combined with NO), NO, and O_2.[15]

c. Direct CI/MS Analysis

Multidrug mixtures have been applied directly to the probe of the CI mass spectrometer in order to identify compounds rapidly by abundant and unique quasimolecular ions. Therefore, this method of direct analysis is useful in cases of drug overdose in which rapid identification is critical. The disadvantages, however, are the inability to differentiate between some pairs of drugs, that relatively large quantities of material are needed, and, of course, that the presence of nondrug components could be falsely interpreted.

d. GC/CI/MS

The CI mass spectrometer is more easily combined than its EI counterpart with the gas chromatograph. The ability of the CI source to handle high gas loads decreases the pressure differential between the two instruments. When reagent gases (H_2O, NO, O_2) that may adversely affect the GC stationary phase are used, they should be introduced as far as possible from the GC interface.

e. Negative Ion Formation

In conventional ion sources, negative ions which are produced are normally trapped. A reversal of the potentials of the ion repeller and accelerator voltage plates can result in the acceleration and focusing of these negative ions.

Many compounds which show a response to ECD can also be determined mass spectrometrically by the resultant negative ions formed which lead to that response. The use of ECD/GC in the analysis of compounds present in low concentrations in highly complex samples is effective as a selective detector. Typical compounds which respond to ECD are halogen-containing drugs, derivatized compounds (with halogen-containing derivatizing agents), substituted phenols, biphenyls, polycyclic aromatics, and heteroaromatics. Additionally, oxygen as a CE gas appears to be particularly useful for negative CI studies on polyaromatic and polyhalogenated compounds. A negative ion mass spectrum can provide direct knowledge of the chemical events occurring under ECD/GC conditions.

The resultant mass spectrum provides far less structural information than its positive ion counterpart. However, in quantitative procedures for selective molecules, electron capture currents can exceed their corresponding positive ion currents by two or three orders of magnitude.

III. DRUGS OF ABUSE

Drugs of abuse are dependence-producing substances which include narcotics, cannabis, depressants, stimulants, and hallucinogens.[19]

A. Narcotics — Opium, Opium Derivatives, and Synthetic Substitutes

1. Opium and Opium Derivatives

The prototypical dependence-producing constituents and derivatives of opium possess a phenanthrene-like chemical structure. These substances include morphine, codeine, thebaine, and heroin. The mass spectra of morphine and related compounds exhibit features desirable for mass spectral analysis because these substances elicit intense molecular ions as well as numerous diagnostic fragment ions.[20-21] Therefore, GC/MS offers the necessary specificity and sensitivity for obtaining an unequivocal determination of morphine; the procedure has also been used for the forensic identification of constituents of opium[22] as well as its metabolites.[23-24] Gas chromatographic procedures of these substances in body fluids sometimes require prior extraction and formation of a volatile derivative by silyation or alkylation.[25-29]

Cone et al.[24] developed a GC/MS assay for eight opium alkaloids in human urine following opium ingestion. Compounds were extracted from urine, converted to their trimethylsilyl (TMS) derivatives, and analyzed by CI mass fragmentography.

In order to provide the needed sensitivity and specificity for analysis of a multicomponent mixture, ions were selected from each component spectra: morphine (m/z 340 and 414), codeine (m/z 282 and 372), *nor*morphine (m/z 326 and 400), *nor*codeine (m/z 268 and 358), thebaine (m/z 312 and 340), oripavine (m/z 370 and 354), papaverine (m/z 340 and 368), noscapine (m/z 220 and 195), and α-isocodeine (internal standard — m/z 372). The method was sensitive to 0.01 μg/mℓ for morphine and codeine and to 0.05 μg/mℓ for the other substances. Selected ion recordings of urinary extracts of standards and "opium eater's" urine shown in Figure 1.

Cone et al.[23] have also described another mass spectrometric assay for codeine, morphine, and the following potential urinary and blood metabolites: *nor*codeine, *nor*morphine, hydrocodone, 6α-hydrocodol, 6β-hydrocodol, *nor*hydrocodol, α-isocodeine, hydromorphone, 6α-hydromorphol, 6β-hydromorphol, and α-isomorphine. The retention times and seven different ions (m/z 372, m/z 430, m/z 268, m/z 326, m/z 374, m/z 358, and m/z 432) were selected for monitoring and determining each compound. The procedure applied to the CI mass spectra of their TMS derivatives. An example is shown in Figure 2.

Jones et al.[30] determined traces of morphine in lung, kidney, and liver tissues from rats 22 days post-withdrawal. The procedure involved monitoring m/z 414 for the pentafluoropropionyl derivative of morphine and m/z 417 for that of the internal standard, [NC^{2}H$_3$]-morphine. The use of capillary columns over packed columns enhanced sensitivity ten times.

Drost et al.[31] described an efficient procedure for the isolation of morphine from serum and cerebrospinal fluid (CSF), followed by selected ion monitoring of the TMS derivatives of morphine and internal standard, [N-C^{2}H$_3$]-morphine. The SIM was performed by CI (ammonia-methane) MS focused on m/z 340 and m/z 343, which corresponded to loss of [(CH$_3$)$_3$-SiOH]$^+$ from the molecular ions of the two substances. Figure 3 shows the CI mass spectra of the TMS derivatives of morphine and the internal standard.

An assay for the more potent semisynthetic narcotic analgesic, oxymorphone, (14-hydroxydihydromorphinone), as well as its metabolites, was developed utilizing GC/MS also in the CI mode.[32] The TMS derivatives of oxymorphone, metabolites, and internal standard (6β-hydrocodol) were quantitated.

Etorphine (6,14-endo-etheno-7-[1-*(R)*-hydroxy-1-methylbutyl]-6,7,8,14-tetrahydrooripavine) is an analgesic with approximately 1000 times the potency of morphine. Its study in biological systems has been difficult since it cannot be detected in urine by many analytical approaches after the administration of pharmacologically effective doses.

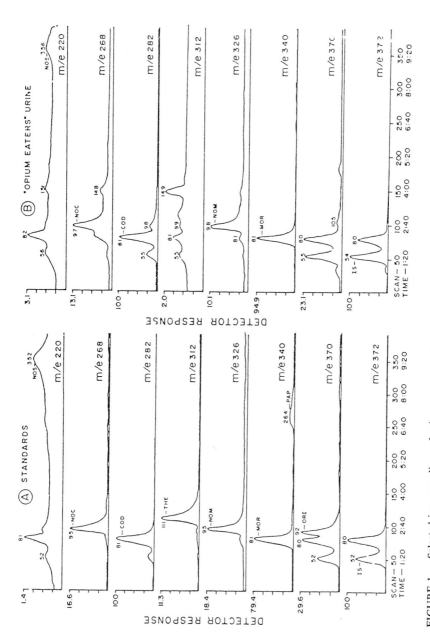

FIGURE 1. Selected ion recordings of urinary extracts of standards (A) and "opium eater's" (B) urine. (From Cone, E. J. et al., *J. Chromatogr.*, 230, 57, 1982. With permission.)

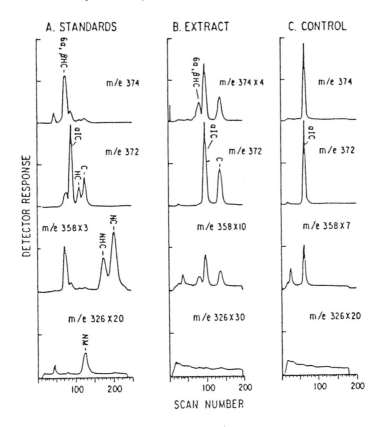

FIGURE 2.   Mass fragmentograms of TMS derivatives of guinea pig urine extracts. (A) Control urine with added standards; (B) 24-hr sample following subcutaneous administration of codeine (15 mg/kg); (C) control urine collected prior to drug administration. (From Cone, E. J. et al., *J. Chromatogr.*, 275, 307, 1983. With permission.)

A sensitive GC/MS assay for the determination of etorphine in urine was developed.[33] The sensitivity of the method was about 5 ng/ml, which compared favorably with a sensitivity of 100 ng/ml by GC analysis. The procedure included repeatedly basifying the urine, extraction into butyl chloride, and reacidification. A final organic extract was derivatized to form the TMS derivative. $^3H_2$-Etorphine was added to control urine and treated by the identical procedure. The mass spectrum of etorphine showed a molecular ion at m/z 483. The spectrum of tritiated etorphine showed a molecular ion at m/z 487($^3H_0$) and ions at m/z 483($^3H_0$) and 485($^3H_1$). Samples were extracted, derivatized, and analyzed on the GC/MS focused at m/z 483 and 487.

Several CI/MS SIM methods are available for monitoring heroin in patients on methadone maintenance. This is accomplished by determining morphine (the heroin metabolite) in blood and urine. Methods are reported that use a stable isotope internal standard — N-trideuteromethyl-morphine — which was derivatized in one case with trifluoroacetic anhydride (TFA)[34] and in the other, with trimethylsilylated with N,O-bis(trimethylsilyl)acetamide.[35] The former case was a study of heroin hydrolysis in blood plasma and therefore also included determination of O^6-monoacetylmorphine with its respective deutero-TFA derivative as internal standard. For the TFA derivative, m/z 364 (protio compound) and m/z 367 (deutero compound) were chosen to be monitored by SIM because they were high molecular weight fragments and each provided a high relative total intensity. The same fragments, m/z 364 and m/z 367, were

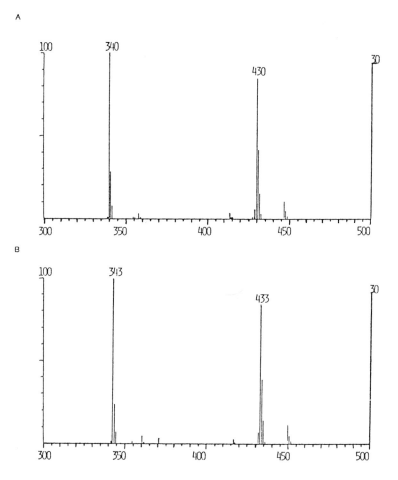

FIGURE 3.   Ammonia-methane CI spectra of TMS derivatives of morphine (A) and internal standard, [N-²H₃]-morphine (B). (From Drost, R. H. et al., *J. Chromatogr.*, 310, 193, 1984. With permission.)

monitored for determination of O^6-monoacetylmorphine. Prior to the GC/MS run of the plasma samples, the standards, trideuteromorphine and trideutero-O^6-monoacetylmorphine, were added to the sample and therefore subjected to identical chemical treatment as the substrate being analyzed.

In the other morphine determination, in which the TMS derivatives were analyzed, four masses were monitored: m/z 340, 343, 414, and 417. Although it was only necessary to monitor one pair of masses, monitoring an additional pair provided convenient corroborative information.

The agonist-antagonist analgesics that are chemical derivatives of opium constituents have similarly been determined by GC/MS. Buprenorphine, a thebaine derivative, is converted in humans to its N-dealkylated metabolite (norbuprenorphine) and conjugated buprenorphine and norbuprenorphine. Urinary extracts were derivatized with pentafluoropropionic anhydride, followed by analysis by GC/MS operating in negative (NCI) and positive CI (PCI) modes.[36] The identity of norbuprenorphine was established by the fragmentation pattern of the NCI mass spectrum. Norbuprenorphine was detected in the urine and feces of subjects receiving buprenorphine by oral, sublingual, and subcutaneous routes of administration. Figure 4 shows the CI negative-ion mass spectrum of norbuprenorphine in urine extract.

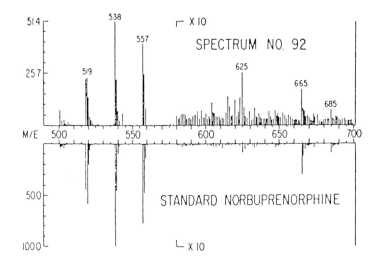

FIGURE 4.   GC-MS CI negative ion mass spectrum of norbuprenorphine in human urine extract following the administration of an oral dose of 40 mg norbuprenorphine. (From Cone, E. J. et al., *Drug Metab. Dispos.*, 12, 577, 1984. With permission.)

Another recent study for the SIM determination of buprenorphine and norbuprenorphine included subjecting the substances to mild acid hydrolysis.[37] This was followed by quantitative loss of methanol and chemical rearrangement of the molecules. The products, after derivatization with pentafluoropropionic anhydride, provided improved sensitivity by GC/MS.

The metabolic disposition of codorphone, another thebaine derivative, which has structural features similar to morphine and the narcotic antagonist, naltrexone, has also been studied by GC/MS.[38] Formation of the trimethylsilyloxime derivatives of the drug and metabolite facilitated analysis by helping to avoid the enolization of each substance that usually occurs. Additionally, an aliquot of the reaction mixture could be injected directly into the gas chromatograph. A number of diagnostic ions formed and aided in identifying codorphone metabolites in complex mixtures.

2. Synthetic Narcotics

Abuse of fentanyl [1-(2-phenethyl)-4-*N-(N*-propionylanilino)piperidine] and its analogs is a current major concern. The potent and totally synthetic narcotic, fentanyl, and its metabolites have been determined by several recent GC/MS procedures.[39,40] The high relative potency of fentanyl makes detection and structural elucidation of its metabolites difficult. In addition, fentanyl does not yield a detectable molecular ion in its mass spectrum; the characteristic base peak was at m/z 245, corresponding to (M − $CH_2C_6H_5$)[+].

Goromaru et al.[39] determined the metabolites of fentanyl, as TMS derivatives, in rat urine by SIM with ($^2H_5$)-fentanyl as internal standard. The metabolites were identified as 4-*N-(N*-propionylanilino)piperidine, 4-*N-(N*-hydroxypropionylanilino)piperidine, 4-*N-(N*-propionylanilino)hydroxypiperidine, 1-(2-phenethyl)-4-*N-(N*-hydroxypropionylanilino)piperidine, and 1-(2-phenethyl)-4-*N-(N*-propionylanilino)hydroxypiperidine. In addition, 4-*N-(N*-propionylanilino)piperidine and 1-(2-phenethyl)-4-*N*-anilinopiperidine were detected by GC/MS and GC (FID) in the plasma of patients treated with fentanyl.[41]

Following a death as a result of drug abuse from self-injection of fentanyl, a method for its GC/MS SIM detection in autopsy specimens was recently reported.[42] Fentanyl concentrations of 4.8, 6.3, and 4.7 $\mu g/\ell$ found in blood, urine, and bile, respectively, were similar to those reported for fatalities from the potent α-methylfentanyl, a clandestinely manufactured fentanyl analog.

Levorphanol (17-methylmorphinan-3-ol) was assayed by GC/MS (NCI) by selected ion monitoring of the strong molecular anion of its pentafluorobenzoyl derivative.[43] A trideuterated analog of levorphanol was used as the internal standard. The pentafluorobenzoyl derivative enhances the sensitivity for detection of phenolic morphines and morphinans analyzed by GC/MS.[44]

Meperidine (pethidine) pharmacokinetics and metabolism have been studied in a variety of species with the aid of GC/MS.[45-47] Meperidine is extensively metabolized by humans and animals. Its major metabolites are *nor*meperidine, 1-methyl-4-phenyl-4-piperidinecarboxylic acid and 4-phenyl-4-piperidinecarboxylic acid. The minor metabolites are the *p*-hydroxy derivative of meperidine, the N-oxide of meperidine, the methoxyhydroxy derivative of meperidine, and the N-hydroxy derivative of *nor*meperidine.

Four new meperidine metabolites were identified by GC/MS in the urine of rats, guinea pigs, rabbits, cats, and dogs.[47] They include 4-ethoxycarbonyl-4-phenyl-1,2,3,4-tetrahydropyridine (dehydro*nor*meperidine), the N-hydroxy dehydro derivative of *nor*meperidine, the dihydroxy derivative of meperidine, and the dihydroxy derivative of *nor*meperidine. For the GC/MS studies, meperidine and its metabolites were extracted from urine, converted to their TMS derivatives, and analyzed by CI/MS. The peaks analyzed included the $(M + H)^+$, $(M + C_2H_5)^+$, and $(M - CH_3)^+$ ions. The disposition of meperidine and the N-demethylated metabolite, *nor*meperidine, in elderly patients was studied by GC/MS in order to determine the effect of age on meperidine analgesia.[45]

Blood and urine samples collected over 24 hr from elderly and younger subjects were compared; drug clearance and plasma concentrations in the two groups did not differ significantly.

Ketobemidone and its N-demethylated metabolite, *nor*ketobemidone, have also been determined in human plasma by GC/MS with SIM procedures.[48-50] The earlier methods utilizing prior diacetylation of the phenol and the amine groups were less favorable since simultaneous determination of the drug and metabolite was difficult. The analysis for simultaneous determination of both ketobemidone and *nor*ketobemidone was facilitated by derivatization with ethylchloroformate.[50] The ions monitored were m/z 319 (molecular ion of the *O*-ethoxycarbonyl derivative of ketobemidone), m/z 321 (loss of 56 amu for the *nor*ketobemidone derivative), and m/z 323 and 325 for the derivatized internal standards, $[^2H_4]$-ketobemidone and $[^2H_4]$-*nor*ketobemidone, respectively.

Finally, methadone and its metabolites in body fluids have also been analyzed by EI/MS[51,52] and CI/MS[13,53] by SIM procedures. Optimum sensitivity has been obtained by application of a deuterated-methadone as internal standard.

CI/MS has permitted more accurate measurement of low methadone levels (5 to 40 mg/day) present in adolescent heroin addicts on methadone maintenance therapy and in newborn infants of mothers who are maintained on methadone. This method also has permitted studies in animals in which plasma methadone levels are lower and turnover rates are much faster than in humans.[13]

The CI/MS/SIM method for monitoring methadone maintenance individuals was an improvement over previous GC analyses that did not provide reliable quantitation for concentrations less than 10 ng/mℓ with high precision. The internal standard was $^2H_5$-methadone, and body fluids analyzed included plasma and urine. Isotope ratios were obtained using isobutane CI and by monitoring the protonated molecular ions at m/z 310 and 315 for methadone and $^2H_5$-methadone, respectively. This procedure was an improvement over EI ion monitoring methods, in which the $(M - 15)^+$ m/z 294 ion was used for quantitation and its relative abundance was low. This procedure used one of the most intense ions in the CI mass spectrum of methadone [m/z 310, $(MH)^+$]. The sensitivity was also about a factor of 15 to 20 better than GC procedures, which also required a larger volume of plasma.[13]

B. Cannabis

Cannabis is the material from the plant (*Cannabis sativa* L.) which is smoked in the form of marijuana cigarettes or is extracted to yield hashish oil. Δ-9-Tetrahydrocannabinol (Δ⁹-THC) is believed to be responsible for most of the characteristic psychoactive effect of cannabis.[19] The accuracy and significance of measurements of Δ⁹-THC and/or its metabolites is of current concern. The reliability of urine screening procedures that are used to determine whether individuals are smoking marijuana depends upon determination of Δ⁹-THC and its metabolites. Rust,[54] reporting on improper marijuana testing procedures, stated that GC/MS is the confirmatory approach for positive Δ⁹-THC identification.

Also at issue is that a positive indication of cannabis in body fluids may result from the passive inhalation of marijuana smoke.[55-56] Although sensitive, specific, and reliable mass spectrometric procedures have been developed, attention must be paid to whether the presence of Δ⁹-THC in body fluids means that marijuana has, in fact, been smoked.

1. Improved Derivatization of Cannabis Constituents for EI/MS Analysis

The use of improved derivatization techniques has provided better gas chromatographic separations of similar substances and has increased mass spectral capabilities in structural elucidation of components in complex mixtures. Components of cannabis extracts when converted to their TMS derivatives have been separated fairly well on a 3% SE-30 column. However, some mono- and dihydroxy components were poorly resolved. Their separation was achieved by variation of the derivatizing agent, i.e., substitution of the TMS group with silyl moieties containing higher alkyl substituents. This resulted in the production of longer retention times for cannabis diols, thus shifting their respective signals away from the other constituents while still retaining good gas chromatographic properties and being stable. The tri-*n*-alkylsilanes included triethyl-, tri-*n*-propyl-, tri-*n*-butyl-, and tri-*n*-hexylsilanes.[57] Derivatives with alkyl groups greater than four carbons required, however, undesirably high elution temperatures. By increasing the molecular weight of the silyl moiety, however, SIM was improved due to lack of background ions at the higher masses. In general, the mass spectral characteristics of the higher alkylsilyl derivatives were similar to the TMS derivatives.

Fresh samples of *Cannabis sativa* L. usually contain cannabinoids in the form of their carboxylic acid derivatives (Δ⁹-tetrahydrocannabinolic acid). These acids cannot be examined directly by GC since they decarboxylate upon heating, even to some extent as the TMS ethers. Preparation of cyclic alkylboronate derivatives (alkyl=methyl or butyl) has been shown to be suitable for GC/MS studies on these types of compounds. In addition, isomeric acids in which the phenolic hydroxyl was *para* to the carboxylic acid have been reported to be present in cannabis. Since they are incapable of forming cyclic alkylboronates, this method of derivatization provided an ideal means of distinguishing between the two isomers.[58] The methylboronates had retention times comparable with those of the TMS derivatives, whereas the *n*-butylboronates had longer retention times. A major advantage over TMS derivatization was the reduction in molecular weight obtained for all of the alkylboronate derivatives as the result of replacing two TMS groups with the relatively low mass boronate moiety. In addition, the mass spectral characteristics observed for the alkylboronate derivatives included positive charge localization away from the boronate ring systems onto the heterocyclic oxygen atom. The spectra of the boronate derivatives exhibited molecular ions as the most abundant in the spectra. Additional derivatives, including *tert*-butyldimethylsilyl (TBDMS), cyclotetramethyleneisopropylsilyl (TMIPS), and cyclotetramethylene- *tert*-butylsilyl (TMTBS) derivatives[59] offered several advantages over TMS, including better separations by GC and structural information by EI/MS. Due to steric crowding, these bulky groups have decreased susceptibility to nucleophilic attack and therefore have greater stability toward hydrolysis than do TMS ethers.

Like TMS ethers, the bulky silyl ethers offer good volatility and thermal stability and some other characteristic properties make them valuable for identifying fragment ion types and deducing fragmentation pathways in MS.

The mass spectra of these ethers are usually dominated by peaks arising from initial siliconium formation, rationalized by elimination of a stable, branched alkyl radical, thus relieving steric crowding in the silyl group. Ions of these derivatives are more prominent in the important high mass region of mass spectra than in those of the TMS ethers.

2. Δ-9-Tetrahydrocannabinol

Δ^9-THC was measured in body fluids by GC/MS by Agurell et al.[60] The procedure included prior liquid chromatographic extraction and EI ionization MS. Rosenthal et al.[61] compared the mass spectral measurements of Δ^9-THC in plasma by EI and CI methods and reported that both techniques could be used to measure concentrations as low as 0.5 ng/ml.

Pirl et al.[62] developed a GC/MS assay for THC in post-mortem blood which was sensitive to 0.5 ng/ml for a sample volume of 15 to 25 ml. The method reported by Harvey et al.[63] for measuring THC (as the TMS derivative) in plasma by monitoring metastable ions proved extremely sensitive, capable of measuring THC in 1 ml of plasma at the 5-pg/ml level. Following a single 0.1-mg/kg dose, THC could be monitored up to 4 days after administration.

3. Metabolites of Δ-9-Tetrahydrocannabinol

A major metabolite of Δ^9-THC is 11-*nor*-Δ^9-tetrahydrocannabinol-9-carboxylic acid; its determination has been reported by RIA, EIA, HPLC, GC/ECD, and GC/MS.[64]

Significant correlations between results of HPLC, GC/ECD, and GC/MS were obtained. All produced lower values than those obtained by either immunologic assay (RIA or EIA). Another study was used to compare several cannabinoid urine assays: GC (FID), RIA, and GC/MS.[65] The GC/MS procedure proved to be the most reliable method, confirming practically all positive results by the other methods.

A recent procedure has been developed for measuring Δ^9-THC in the plasma of heavy and light cannabis users up to 72 hr after administration.[66] The substance was deuterium labeled and was administered by smoking as well as by intravenous injection. The tetrahydrocannabinol administered was [$^2H_3$]-THC and the internal standard was [$^2H_7$]-THC. The body fluid extract was derivatized with silylating agents to form TMS-[$^2H_3$]-THC and TMS-[$^2H_7$]-THC. SIM of the molecular ion of each was performed at m/z 389 and m/z 393, respectively.

A rapid simplified GC/MS method utilizing EI ionization for determining 11-*nor*-Δ^9-tetrahydrocannabinol-9-carboxylic acid from urine has been developed for the routine determination of large numbers of samples.[67] The metabolite is detected as the

TMS derivative; its detection limit is below the 1 ng/mℓ level with [^{2}H$_3$]-11-*nor*-Δ^9-tetrahydrocannabinol-9-carboxylic acid as internal standard.

Another rapid GC/MS method has also been developed for determining Δ^9-THC and two of its metabolites, 11-hydroxy-Δ^9-tetrahydrocannabinol and 11-*nor*-Δ^9-tetrahydrocannabinol-9-carboxylic acid.[68] The procedure, involving the use of capillary column GC, NCI/MS, and deuterium-labeled analogs of each cannabinoid as internal standards, allowed analysis of a single 1-mℓ sample of body fluid. The key difference between this method and other procedures was the derivatization with TFA anhydride and the obtaining of two extracts: a neutral fraction containing Δ^9-THC and 11-hydroxy-Δ^9-tetrahydrocannabinol, and an acid fraction containing 11-hydroxy-Δ^9-tetrahydrocannabinol-9-carboxylic acid. Each derivatized substance produced abundant and diagnostic molecular anions or fragments thereof which were quantified and which reportedly produced greater sensitivity and selectively than that obtained by PCI or EI. The limits of reliable measurement were 0.2 ng/mℓ of Δ^9-THC, 0.5 ng/mℓ for 11-hydroxy-Δ^9-tetrahydrocannabinol, and 0.1 ng/mℓ for 11-*nor*-Δ^9-tetrahydrocannabinol-9-carboxylic acid. Figure 5 shows the negative ion CI mass spectra of the TFA derivatives of the THC, HO-THC, and COOH-THC methyl ester.

Other quantitative and qualitative procedures for determination of the carboxy-THC metabolite in body fluids by GC/MS have been reported.[69,70] Harvey et al.[71] have recently investigated the in vivo metabolism of Δ^9-THC in rabbits; 16 metabolites were identified by the GC/MS method based on metastable ion monitoring.[63] Its metabolism is complex, showing considerable species variability. Major sites of metabolism are the allylic positions 6 and 7, and the aliphatic carbons of the pentyl side chain. Resulting hydroxy compounds are readily oxidized to aldehydes, ketones, and acids, as well as their glucuronide conjugates.

Mason et al.[55] and Wethe et al.[56] have reported Δ^9-THC and its metabolites in body fluids of subjects known to be passively exposed to marijuana smoke. Mason et al.[55] detected Δ^9-THC and the carboxy-THC metabolite in the plasma of a subject passively exposed to smoke for a 1-hr period (exaggerated conditions). Wethe et al.[56] described the GC/MS determination of blood from a subject exposed to smoking the equivalent of 90 mg Δ^9-THC during a 30-min period. The Δ^9-THC levels in all blood samples were 4 to 20 pmol/mℓ.

C. Depressants — Barbituric Acid Derivatives and the Benzodiazepines
1. Barbiturates

Barbiturates are frequently prescribed to induce sedation or sleep. Recent mass spectrometric assays for phenobarbital, methohexital, and mephobarbital in body fluids have been reported. The development of the assay is often for therapeutic monitoring for a drug when administered alone or in combination, as well as for studying the pharmacokinetics and biotransformation of drugs during therapy.[72] The barbiturates are often subject to forensic analysis because of their involvement in suicides and accidental overdoses.

Several methods have been described which are suitable for the simultaneous determination of phenobarbital and metabolites or other drugs.[73-76] A GC/MS method for quantitation of phenobarbital and p-hydroxyphenobarbital in serum and urine was reported recently by Van Langenhove et al.[72] The method was useful for determining each substance in a wide concentration range, both in serum (0.1 to 30.0 μg/mℓ) and in urine (1.0 to 50.0 μg/mℓ). The internal standards containing stable isotopes included the following: 5-ethyl-5(2,3,4,5-tetradeuterophenyl)-2-(^{13}C)-barbituric acid, and 5-ethyl-5-(4-hydroxy-3,5-dideuterophenyl)-2-(^{13}C)-1,3-(^{15}N$_2$)-barbituric acid for phenobarbital and p-hydroxyphenobarbital, respectively. The substances were derivatized to

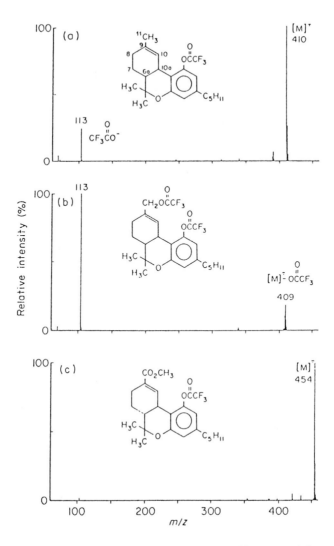

FIGURE 5. Negative-ion CI mass spectra of trifluoroacetate derivatives of THC (a), HO-THC (b), and COOH-THC methyl ester (c). Methane was the CI reagent gas. (From Foltz, R. L., *Biomed. Mass Spectrom.*, 10, 316, 1983. With permission.)

their methylated products with methyl iodide. The mass ranges of m/z 231 to 236 and m/z 289 to 296 were then scanned for the derivatized phenobarbital and p-hydroxyphenobarbital, respectively. The mass ranges encompass major fragments at $(M - 28)^+$ of methylated phenobarbital (m/z 232,235,237) and $(M)^+$ for methylated p-hydroxyphenobarbital (m/z 290,293,295).

Another GC/MS procedure has been reported for the simultaneous determination of phenobarbital, p-hydroxyphenobarbital, and a third metabolite, phenobarbital-N-glucoside, in urine, along with the stable isotope-labeled internal standard, ^{15}N-phenobarbital.[77] The urine extract was methylated, silylated, and analyzed by CI/MS. The assay method was applied to monitoring patients' urinary excretion of phenobarbital and its metabolites for 20 days after administration of the drug.

Hooper et al.[74] reported on the sensitive and simultaneous assay of methylphenobarbital (mephobarbital) and its metabolite, phenobarbital, in plasma. Previously, separate CI assays for each substance had been developed; the assays did not allow, how-

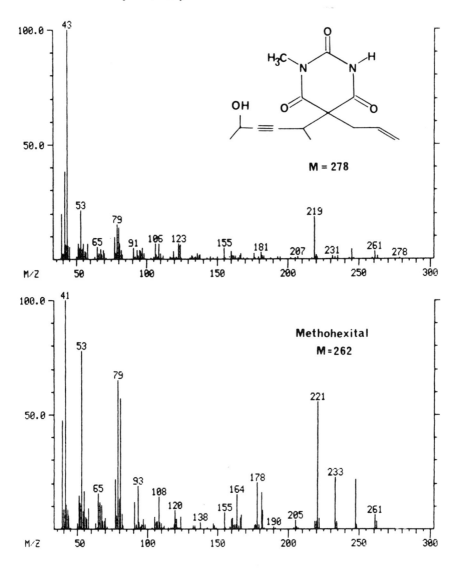

FIGURE 6.   EI mass spectra of methohexital (lower spectrum) and its metabolite 4'-hydroxy-methohexital (upper spectrum). (From Heusler, H. et al., *J. Chromatogr.*, 226, 403, 1981. With permission.)

ever, for their simultaneous determination.[78,79] The internal standard selected was 5-ethyl-5-(4-methylphenyl)barbituric acid. Substances were converted to the propyl derivatives for which the base peaks were monitored.

The ultra-short acting barbiturate, methohexital, has been measured in concentrations less than 50 ng/mℓ in plasma or whole blood samples of humans or rats.[80] Its major metabolite, 4'-hydroxymethohexital (1-methyl-5-allyl-5-(1'-methyl-4'-hydroxy-pentyn-2'-yl)barbituric acid), was identified with the aid of proton NMR spectroscopy and GC/MS. Its characteristic m/z values were (M)⁺ = m/z 278, (loss of the hydroxylated 2-pentyne group) = m/z 181, and (loss of the allyl radical and a molecule of water) = m/z 219 (Figure 6).

2. Benzodiazepines

The benzodiazepines are depressants used for the treatment of anxiety, tension, and muscle spasms. They are marketed as mild or minor tranquilizers, sedatives, hypnotics, or anticonvulsants. Their margin of safety is generally considered greater than that of other depressants. To some extent, the benzodiazepines have replaced the barbiturates in medical treatment.

The benzodiazepines have been analyzed by NCI/MS on numerous occasions.[81-83] Diazepam (7-chloro-1,3-dihydro-1-methyl-5-phenyl-2*H*-1,4-benzodiazepin-2-one), the prototype of the benzodiazepines, and its major metabolite, *nor* diazepam, have shown prominent $(M - 1)^-$ molecular anions in their NCI mass spectra.[84] This characteristic has been attributed to its reaction in the ion source of the mass spectrometer with trace amounts of oxygen; in a rebuilt GC/MS interface, the $(M - 1)^-$ anions disappeared, being replaced by the molecular anions of diazepam and *nor* diazepam.

NCI/MS has recently been used to aid discovery of a new metabolite from another common benzodiazepine, flurazepam;[85] it has been identified as 7-chloro-5-(2'-fluorophenyl)-2,3-dihydro-2-oxo-1,4-benzodiazepine-acetaldehyde. The metabolite is believed to contribute to some of the pharmacological effects of flurazepam. Negative ion CI mass spectra of plasma extracts of humans and cats fed with flurazepam are shown in Figure 7.

Midazolam is a newly marketed benzodiazepine pharmacologically similar to diazepam, but with a much shorter half-life. Its indicated application is either alone or in combination with other substances in anesthesia. In order to establish therapeutic and toxic levels, midazolam was assayed in serum and urine by GC/MS, with flurazepam used as internal standard.[86] The metabolite, 1-hydroxymethylmidazolam, was determined by liquid chromatography as it did not chromatograph well by GC. Another metabolite, desmethylmidazolam, was determined by GC. A subsequent study by Rubio et al.[87] was developed to measure midazolam and its metabolites as TMS derivatives in human plasma by GC/MS (EC/NCI) SIM. For the three substances, the limit of quantitation was 1 ng/m*l*; however, no measurable levels of desmethylmidazolam were attained.

A sensitive GC, ammonia-^{15}N isotope dilution assay was developed to measure the benzodiazepine clonazepam and its 7-amino metabolite in blood or plasma.[88] The method was used to measure both compounds in the blood of 1 subject administered a single 2-mg dose of clonazepam, and in the plasma of 13 subjects on a clonazepam oral dosing regimen. The assay developed involves a simple extraction of plasma or blood, followed by GC/MS analysis of the residue after evaporation of the extraction solvent. Specificity was obtained by monitoring the MH$^+$ ions of clonazepam and its 7-amino metabolite, generated by ammonia CI. Assay accuracy was insured by the use of stable isotope analogs of clonazepam and the metabolite as internal standards. The ions monitored were m/z 286 (MH$^+$ of 7-amino metabolite), m/z 287 (MH$^+$ of ^{15}N-7 amino metabolite), m/z 316 (MH$^+$ of clonazepam), and m/z 317 (MH$^+$ of ^{15}N-clonazepam). The limit of detection of the method, 1 ng/m*l* for clonazepam and 2 ng/m*l* for the 7-amino metabolite, was sufficient for measuring clonazepam and its metabolite in most subjects on the chronic dosing regimen, and in many subjects following administration of a single dose of clonazepam.

D. Stimulants — Amphetamine, Amphetamine Analogs, and Cocaine
1. Amphetamine and Amphetamine Analogs

The abuse of amphetamine, methamphetamine, and their analogs is widespread. Improved methods for their separation, identification, and quantitation are needed. Several recent procedures using GC/MS have identified these drugs in body fluids.[89-93] CI/MS has also been used to quantitate amphetamine and methamphetamine in blood and urine.[91,92] In addition, the use of derivatizing agents has been used to aid in the analysis of the amphetamine analogs.[94-96]

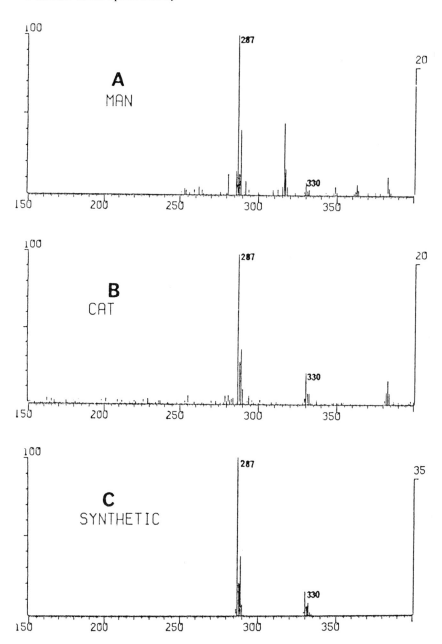

FIGURE 7.   Negative ion CI mass spectra of plasma extracts of humans and cats fed with flurazepam. (From Garland, W. A. et al., *Drug Metab. Dispos.*, 11, 70, 1983. With permission.)

The *N*-trifluoroacetyl (*N*-TFA) derivative improves the gas chromatographic properties of amphetamine and its analogs.[90,94] It is produced by reaction of the amine with either TFA anhydride or *N*-methyl-*bis*-(trifluoroacetamide). The *N*-TFA derivative offers advantages when used in conjunction with GC/MS studies for qualitative and quantitative analysis. It yields a more complex, but unique, mass spectrum.[97,98] Illicit drug samples, although not originating from body fluids, have also been converted into their *N*-TFA derivatives and analyzed by GC/MS.[93]

Foltz et al.[90] reported on formation and analysis of the *N*-TFA derivatives of the following stimulants: β-phenethylamine, amphetamine, phentermine, methamphetamine, mephentermine, p-chlorphentermine, o-chlorphentermine, p-hydroxyamphetamine, ephedrine, and norephedrine. All gave $(M + H)^+$ peaks as base peaks when analyzed in the underivatized form by CI/MS.

As the N-TFA derivatives, all except ephedrine and norephedrine gave the corresponding $(M + H)^+$ ion in their CI mass spectra. More recently, the N-TFA derivatives of amphetamine, methamphetamine, and some analogs were prepared by an on-column derivatization procedure, applicable for EI, CI, direct probe, and GC approaches.[89]

Coutts et al.[99] described the EI mass spectra of various amines derivatized with either TFA anhydride or pentafluoropropionic anhydride. The amines included amphetamine, ephedrine, p-hydroxyamphetamine, norephedrine, normetanephrine, norpseudoephedrine, and phentermine. Delbeke et al.[100] reported on the derivatization with pentafluorobenzoyl chloride of primary and secondary amines in body fluid samples followed by analysis of their EI mass spectra. Some components yielded weak molecular ions in their mass spectra; nevertheless, the presence of diagnostic ions, such as the stable pentafluorobenzyloxonium ion at m/z 195, allowed confirmation of monobenzoylation of all substances. Derivatives of amines reported included phentermine, chlorphentermine, amphetamine, methamphetamine, mephentermine, N-ethylamphetamine, fenfluramine, methylphenidate, fencamfamine, and phenmetrazine.

Tas et al.[101] reported on the derivatization of catecholamines (dopamine, norepinephrine, and epinephrine) and their 3-O-methylated metabolites (3-methoxytyramine, normetanephrine, and metanephrine) with pentafluoropropionic anhydride. The pentafluoropropionyl derivatives showed good gas chromatographic properties.

Ehrhardt and Schwartz[102] reported on the GC/MS assay of the corresponding benzylic-O-ethyl derivatives of catecholamines in plasma.

Ohki et al.[103] have also reported on the pentafluoropropionyl derivatives of urinary phenolic amines, di- and polyamines, and their GC/MS analyses.

2. Cocaine

Cocaine has become an increasingly popular drug of abuse in the developed countries. Therefore, advanced screening methods that can rapidly and reliably identify cocaine from large numbers of urine samples are needed. Hsu et al.[104] has reported a rapid method for detection of cocaine and its major metabolites by thin-layer chromatography (TLC) with confirmation only by GC. The level of cocaine and its metabolites in biological materials is often, however, extremely low, requiring mass spectrometric methods for determination. Several GC/MS SIM analytical procedures have been reported.[105-108] They involved EI and CI methods. GC/MS by SIM in the CI mode offered greater sensitivity and selectivity than EI, provided that efficient extraction and prior clean-up procedures were used.

Cocaine is metabolized rapidly — primarily to ecgonine, ecgonine methyl ester, and benzoylecgonine. The levels of cocaine, benzoylecgonine, and ecgonine methyl ester were measured in the urine of dogs and rabbits by GC/MS SIM CI after subcutaneous injection of the drug.[109] Urinary ecgonine methyl ester could be identified after 72 hr in experimental dogs. Both metabolites were enzymatic products of cocaine as opposed to spontaneous hydrolysis products.[110] The GC/MS SIM CI (isobutane reagent gas) method was sensitive to 1 ng/mℓ for cocaine and benzoylecgonine and 10 ng/mℓ for ecgonine methyl ester for a 10-mℓ urine sample. Internal standards were [²H₅]-cocaine, [²H₅]-benzoylecgonine, and lidocaine. Other GC/MS methods have used [N-C²H₃]-cocaine and [N-C²H₃]-benzoylecgonine as internal standards.[106,108] A principal part of the procedure which contributed to the high sensitivity of analysis was the elimination of interfering background, attributed to prior treatment of biological specimens with Extrelut®, a diatomaceous earth.

E. Phencyclidine (PCP) — A Hallucinogen

Phencyclidine (PCP) is a hallucinogen which has been described as presenting greater risks to the user than other drugs of abuse. Like other hallucinogens, PCP distorts the perception of reality. It has been responsible for producing violence, hostility, and schizophrenic-like behavior.[19]

A procedure was developed for identifying and measuring PCP in blood with internal standard, 1-(1-phenyl-[$^2H_5$]-cyclohexyl) piperidine (pentadeuterophencyclidine).[111] A known quantity of this internal standard was added to a measured amount of blood, and to aqueous PCP solutions of appropriate concentrations. Each sample was then carried through a three-step separation. The resulting extract was taken to dryness, reconstituted, and two aliquots subjected to analysis, one for quantitation and the other for positive identification. The PCP was quantitated by monitoring the ions at m/z 205 ([$^2H_5$]-phencyclidine) and 200 (phencyclidine). The concentration of PCP was calculated by normalizing the area of the response at m/z 205 to 100%. Applying the factor obtained to the area of the response at m/z 200 yielded the concentration of PCP in micrograms per liter. A second injection of a similar volume was used to confirm the presence of PCP. The ions at m/z 243 (molecular ion), 242, 200, and 186 were monitored and the calculated ratios of the responses compared to that of a standard sample.

The metabolism of PCP has been of considerable interest because some of the effects attributed to the drug itself result from metabolites of unknown structure that may have a pharmacological profile distinct from that of the parent substance.[112] Attempts to explain some of the prolonged neurotoxic effects of PCP have promoted metabolism studies. GC/MS data have been utilized to identify 1-(1-phenyl-4-hydroxycyclohexyl)-4-hydroxypiperidine and 1-(1-hydroxyphenylcyclohexyl)piperidine as PCP metabolites.[113]

Kuhnert et al.[114] developed a simple, rapid, and sensitive GC/MS SIM method for analysis of PCP and hydroxylated PCP metabolites with analogous deuterated internal standards. The metabolites, 4-phenyl-4-piperidino-cyclohexanol (PPC) and 1-(1-phenylcyclohexyl)-4-hydroxypiperidine (PCHP), have been reported as pharmacologically active.[115] The ions selected for monitoring were m/z 96, 200, and 288 for the TMS derivatives of the metabolite and internal standards. Samples containing less than 200 pg/ml PCP, 2 ng/ml PPC, and 1 ng/ml PCHP in urine could be quantitated by the method.

One of the significant findings relating to the metabolism of PCP and its long-term toxic effects is the metabolism-dependent irreversible binding of PCP to cellular proteins in microsomal preparations of various tissues.[116] Incubation of PCP with rabbit liver microsomes and Na^{14}CN resulted in the metabolically dependent formation of a ^{14}C-labeled cyano adduct, which was identified with the aid of HPLC/CI and GC/EI/MS analysis as 1-(1-phenylcyclohexyl)-2-cyanopiperidine.

PCP as a drug of abuse is ingested orally and smoked on cigarettes, which are sometimes composed of parsley. When smoked, substances administered into the subject include PCP as well as decomposition products, 1-phenylcyclohexene and piperidine. 1-Phenylcyclohexene and its metabolites have been found in the plasma and urine of humans and investigated in the rat in vitro by Cook et al.[117]

IV. DRUG OVERDOSE

Due to their widespread use in the treatment of depression, the tricyclic antidepressants were the fourth most common cause of substance overdose seen in hospital emergency rooms in the U.S. in 1982 and the third most common cause of drug-related death, after alcohol-drug combinations and heroin.[118] The four most widely used tricyclic antidepressants are amitriptyline and imipramine and their N-demethylated analogs (and metabolites), nortriptyline and desipramine. Others are doxepin and loxapine. The use of GC/MS SIM has provided the necessary specificity for pharmacokinetic studies and clinical evaluation of these drugs in a sufficiently sensitive range (5 to 10 ng/ml).[119,120]

Ishida et al.[121] have developed a GC/MS/EI SIM method for determination of amitriptyline and several metabolites, including nortriptyline, 10-hydroxyamitriptyline, and 10-hydroxynortriptyline. Better sensitivities were obtained by a GC/MS/CI method; however, this method did not display sufficient linearity in the standard curves of the metabolites.[122] Craig et al.[123] have developed a GC/MS/EI SIM assay for the simultaneous quantitation of imipramine and its metabolite, desipramine, at the nanogram level in a single gas chromatographic peak. [^{2}H$_4$]-Imipramine was used as the internal standard and quantitation was achieved by SIM of the molecular ions and computing ratios of the respective ion currents.

Davis et al.[124] developed a GC/MS method to analyze doxepin and its major metabolite, desmethyldoxepin, with the necessary precision, selectivity, low sample volume (1 ml), and sensitivity of less than 1 ng/ml. Amitriptyline and nortriptyline were used as internal standards. Amoxapine (2-chloro-11-(1-piperazinyl)dibenz[b,f][1,4]-oxazepine), another tricyclic antidepressant, has been studied in several human overdose incidents wherein it elicited seizure and cardiotoxic effects.[125] Its major metabolites, 7-hydroxy- and 8-hydroxyamoxapine, which are pharmacologically active, were determined in urine and serum by GC/MS analysis utilizing the EI mode. Identification of each was confirmed by comparison of reference spectra against unknowns.

Maurer and Pfleger[126-129] have published a series of papers dealing with development of screening procedures for detection of various drugs and their metabolites in urine using a computerized GC/MS system. Procedures such as these are necessary in analytical toxicology in order to rapidly and accurately diagnose drug overdoses from the tricyclic antidepressants as well as other psychotropic and addictive substances. Retention indexes determined by GC procedures provide preliminary indications of the identity of the substance. The mass fragmentogram with eight specific masses scanned allows the detection of 21 antidepressants and their metabolites.[127] Other drug categories for which comparable screening procedures were developed are the phenothiazines[128] and the benzodiazepines.[126]

Two additional substances which possess neuroleptic properties and have been shown to be clinically useful in treating psychoses have been assayed in plasma by various mass spectrometric techniques in order to better define their pharmacokinetics. Melperone (1-(4-fluorophenyl)-4-(4-methyl-1-piperidinyl)-1-butanone), an experimental butyrophenone, was assayed using a combined GC/MS/EI SIM technique with a sensitivity in the low nanogram range using 2 ml of plasma.[130] The chemical structural features of its metabolites were indicated by diagnostic fragment ions. Similarly, flutroline as its TMS derivative was assayed at levels as low as 3 ng/ml in plasma using capillary GC/MS SIM.[131] The internal standard was a chemical homolog of flutroline. The same ion, m/z 266, was monitored for both substances and therefore allowed the instrumental parameters to be set for optimum sensitivity and reproducibility.

Other drugs which have been involved in overdoses and necessitated analysis by mass fragmentography have been reported recently by Semple,[132] Stajic et al.,[133] Ervik et al.,[134] Anderson et al.,[135] and Swahn et al.[136]

Zimelidine, an experimental drug chemically similar to pheniramine, was implicated in a suicide from a gunshot wound. Concentrations of the drug and major metabolite were determined in blood, urine, bile and liver, and brain tissue by capillary GC with a nitrogen-phosphorus detector and confirmed by GC/MS. Death due to drug overdose was ruled out.[132]

Baclofen, β-(p-chlorophenyl-γ-aminobutyric acid), used for treatment of muscle spasticity in patients with multiple sclerosis or central nervous system disorders, was analyzed by GC/MS up to 60 hr after intoxication by 1.23 g.[135] Baclofen from plasma and urine was derivatized with pentafluoropropionic anhydride and pentafluoropropanol, applied to the GC/MS for SIM analysis, and pharmacokinetically evaluated. Minimal pharmacokinetic information in patients had been available owing to difficulties in analyzing baclofen in plasma or urine except by GC/MS[135,136] or GC/ECD.[137]

Overdoses due to ingestion of metoprolol, a β-adrenergic blocking agent used for treating hypertension, have been described by numerous authors.[133,134,138-141] In deaths from metoprolol ingestion, post-mortem toxicological findings have relied on HPLC and GC/MS (CI and EI) to identify the substance in blood, urine, vitreous humor, pleural fluid, bile, kidney, and liver.[133] Authentic standards of metoprolol confirmed the identification of the drug with matching mass spectra and GC retention times. Figure 8 shows the EI and CI mass spectra of metoprolol. Although originally considered as a safe nonbarbiturate sedative-hypnotic, the dependence-producing ethchlorvynol (5-chloro-3-ethylpent-1-yn-4-en-3-ol) (Placidyl®) has frequently been implicated in intoxications and poisonings. Common techniques for analyzing ethchlorvynol by colorimetry and GC have been reported.[142] Horwitz et al.[143] reported the isolation from urine of a hydroxylated metabolite of ethchlorvynol. The identification of the metabolite, 1-chloro-3-ethylpent-1-en-3,4-diol, in urine by GC/MS (Figure 9) can be considered as positive identification for ethchlorvynol.

V. TOXIC SUBSTANCES

Many substances, including environmental pollutants, ingredients in drugs or foods, industrial chemicals, herbicides, and insecticides, are toxic to exposed individuals. Some recent mass spectrometric studies have been useful in studying the pharmacokinetic disposition and metabolism of these substances. Often the metabolite has been found to be responsible for the toxicity of a particular chemical. Construction activities at chemical landfill sites may cause the release of chemical vapors from the sites and lead to exposure of workers and local residents. Analysis of their blood and urine can be used to assess the extent of human exposure to the chemicals.

Yost et al.[144] compared the utility of several mass spectrometric techniques for the screening of chlorinated organic compounds in human blood serum and urine. The objective of the study was to evaluate the potential of tandem mass spectrometry (MS/MS) for rapid screening of the compounds with little or no chromatographic separation. Hexachlorobenzene and 2,4,5-trichlorophenol were the compounds chosen for study as representing two major classes of chemicals found in landfills.

Previous techniques for trace analysis of these chemicals have involved GC/MS. The procedure of Wright et al.[146] lacked the speed necessary for rapid screening of large numbers of samples as would be desired for monitoring workers and area residents. Extensive sample clean-up, required prior to analysis, was very time consuming. Additional work by Brotherton and Yost[145] showed the ability of MS/MS to perform analyses rapidly and with minimal sample clean-up for drug screening in serum and urine.

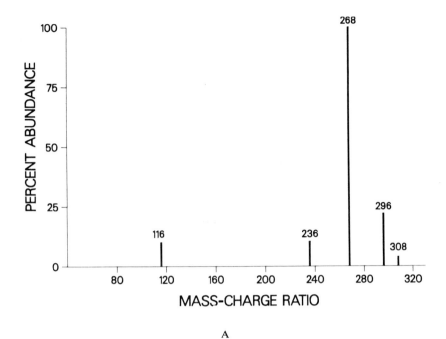

A

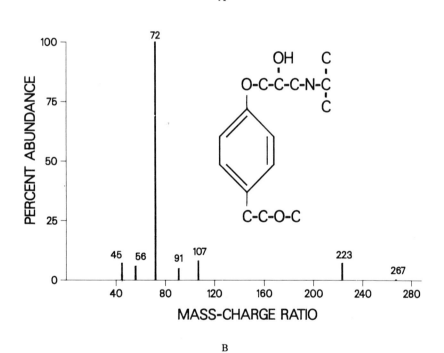

B

FIGURE 8.   (A) EI and (B) CI mass spectra of metoprolol. (From Stajic, M. et al., *J. Anal. Toxicol.*, 8, 228, 1984. With permission.)

Robert and Hagardorn[147] have developed a sensitive quantitative procedure utilizing capillary GC/MS to analyze the chemical pollutant, 2,4-dinitrophenol, the EI mass spectrum of which is shown in Figure 10. This has aided in assessing the contribution of pharmacokinetic factors to its toxicity. The disposition of 2,4-dinitrophenol in serum, liver, and kidney tissues in the mouse was studied. After prior solvent extrac-

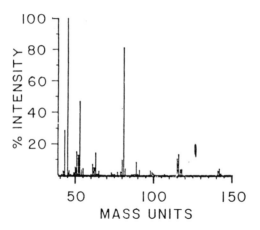

FIGURE 9.   EI mass spectrum of 1-chloro-3-ethyl-
pent-1-en-3,4-diol, a metabolite of ethchlorvynol.
(From Bridges, R. R. et al., *J. Anal. Toxicol.*, 8,
263, 1984. With permission.)

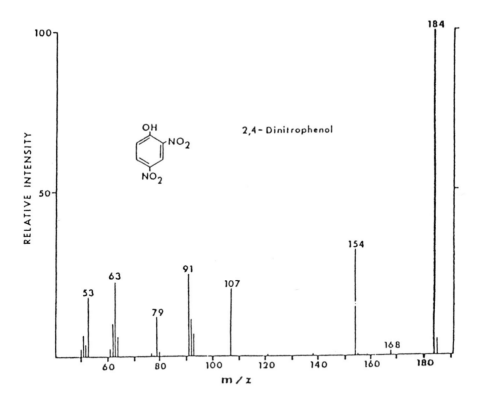

FIGURE 10.   EI mass spectrum of 2,4-dinitrophenol. (From Robert, T. A. et al., *J. Chroma-togr.*, 276, 77, 1983. With permission.)

tion from biological specimens, quantitation was performed by GC/MS SIM of the base peak (also molecular ion) or m/z 184 for 2,4-dinitrophenol and m/z 187 for the internal standard, 3,5,6-[²H₃]-2,4-dinitrophenol. Similar half-lives for absorption, distribution, and elimination (except in kidney) were observed. The primary differences among the serum, liver, and kidney kinetics were the maximum concentrations attained in each and the slow rate of kidney elimination.

The determination of nitrite in blood may need to be performed in cases of intoxication by nitrite-containing drugs or ingestion of food containing nitrite or nitrate. Tanaka et al.[148] monitored the in vivo disappearance of nitrite in the blood of mice by following a single oral dose of $^{15}NO_2$ or $^{15}NO_3$ by GC/MS SIM. $^{15}NO_2$ was derivatized to [^{15}N]-tetrazolophthalazine, and the pentafluorobenzoyl ester of 2,4-dinitro-6-*sec*-butylphenol was the internal standard.

4-Nitroaniline, an industrial intermediate used in the synthesis of many pharmaceutical, veterinary, agricultural, and other consumer products, induces a number of toxicological effects, such as headache, nausea, stupor, cyanosis, jaundice, anemia, and methemoglobinemia leading to anoxia. Anderson et al.[149] studied the in vitro metabolism of 4-nitroaniline using rat liver microsomes to determine if a metabolite would be responsible for any of these effects. Mass fragmentography was used in conjunction with proton NMR spectroscopy to aid in identifying 2-hydroxy-4-nitroaniline as a microsomal metabolite. The substance was extracted with ethyl acetate and isolated by multiple HPLC separations. Its mechanisms of formation involving incorporation of ^{18}O and deuterium exchanges were similarly investigated by mass fragmentography.

The industrial chemicals bromobenzene and *o*-bromophenol both have been shown by similar HPLC retention time and mass spectra fragmentation patterns to metabolize in the rat to 2-bromohydroquinone.[150] Nephrotoxicity observed after administration of the former substances has been attributed to formation of the latter substance. Mass spectra of authentic 2-bromohydroquinone and unknown metabolites of bromobenzene and *o*-bromophenol were compared. The major ions in both spectra were m/z 188, m/z 190, (M$^+$), m/z 108 (M − HBr)$^+$, and m/z 80 (M − HBr − CO)$^+$.

Aromatic hydrocarbons, which include drugs, pesticides, and environmental contaminants, are often metabolized enzymatically to toxic reactive epoxide intermediates. Monks et al.[151] demonstrated the implied formation of bromobenzene-2,3-oxide by identifying bromobenzene-2,3-dihydrodiol as another bromobenzene metabolite. Identification of the chemical structure of the isolated metabolite was accomplished with the aid of mass spectrometric analysis.

2,4-Dichlorophenoxyacetic acid and its esters are used as systemic herbicides for the control of broadleaf weeds in cereal crops, pastures, and urban areas. 2-Ethylhexyl-2,4-dichlorophenoxyacetate was selected for pharmacokinetic evaluation to see if it rapidly hydrolyzed to the parent substance after oral administration and was therefore comparably toxic. The ester was determined by GC/MS SIM used in the EI mode for concentrations greater than 1 μg/mℓ.[152] Ions at m/z 220 and m/z 222 were monitored since they represent the largest fragment ions that yield a good chromatographic response and elicit the identifiable chlorine isotope ratio. GC/ECD analysis supplied a 25-fold increase in sensitivity but had a limited linear range.

Oryzalin (the formulation of 3,5-dinitro-N^4, N^4-dipropylsulfanilamide) is a broad-spectrum and selective herbicide. It may pose a potential hazard to aquatic fauna if large amounts are released into the environment. The chemical has also been implicated as a cause of human birth defects. Bardalaye et al.[153] reported on the derivatization of Oryzalin with methyl iodide and dimethylsulfinyl anion followed by examination of GC/MS EI and CI spectra. The detection limit of the di-*N*-methyl derivative of Oryzalin is 0.05 ng by ECD and 1 ng by NPD. The resulting derivative, N^1, N^1-dimethyl-3,5-dinitro-N^4, N^4-dipropylsulfanilamide, gave characteristic fragments in its EI mass spectrum:

$$[M]^+ = m/z\ 374; \quad [M - C_2H_5]^+ = m/z\ 345; \quad [M - OH]^+ = m/z\ 357;$$
$$[M - CH_3NO]^+ \text{ or } [M - C_2H_5O]^+ = 329;$$
$$[M - (CH_3CH_2 + CH_3-CH{=}CH_2)]^+ = m/z\ 303.$$

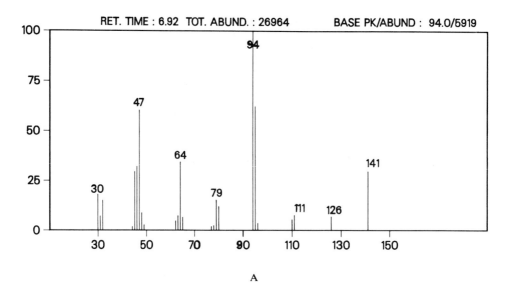

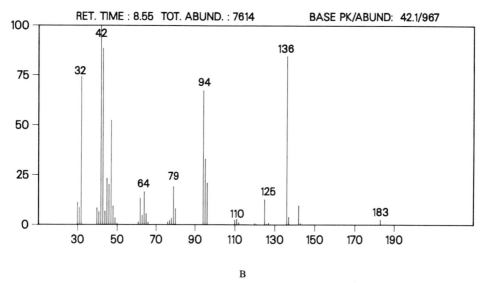

FIGURE 11.   EI mass spectra of acephate (A) and methamidophos (B). (From Singh, A. K., *J. Chromatogr.*, 301, 465, 1984. With permission.)

Additional confirmation of the structure was derived from the methane CI mass spectrum:

$$[M + 1]^+ = m/z \ 375; \ [M + 29]^+ = m/z \ 403; \ \text{and} \ [M + 41]^+ = m/z \ 415$$

The ease of quantitative formation of the di-*N*-methyl derivative of Oryzalin as well as the formation of base peaks $[M - 29]^+$ in the EI mass spectrum and $(M + 1)^+$ in the CI mass spectrum provide potential application for detection of Oryzalin by SIM in low concentrations and in complex mixtures.

GC/MS has been used to unambiguously establish the identity of organophosphorus insecticides. Cairns et al.[154] described a GC/MS technique to identify isofenphos(1-methylethyl-2-[[ethoxy(1-methylethyl)amino]-phosphinothioyloxy]benzoate) as a trace level residue on alfalfa pellets. The work was expanded so as to mass spectrometrically characterize its analogs and metabolites. CI procedures with deuterated methane and ammonia as reagent gases aided in delineating protonation steps and determining exchangeable hydrogens in the overall fragmentation pathways of these compounds.

Acephate is a water-soluble, broad-spectrum organophosphorus insecticide used in the control of agricultural and forestry pests. Its use as an insecticide in the U.S. has increased rapidly due to its low toxicity in mammals (LD_{50} = 900 mg/kg). However, it has been proposed that acephate has the potential of metabolizing in plant and animals to methamidophos, the latter of which is 40 to 50 times more toxic (LD_{50} = 20 mg/kg). Singh[155] reported a rapid GC/MS SIM EI method for the analysis of acephate and methamidophos in biological samples. Figure 11 shows the mass spectra of the two compounds. Ions selected for monitoring acephate were m/z 94 and m/z 136 (retention time = 8.5 min) and for methamidophos, m/z 94 (retention time = 6.9 min). After 24 and 48 hr, 2 ± 0.8 and 5 ± 1.0 pmol/mℓ, respectively, of methamidophos were reported present in the acephate-exposed blood samples.

REFERENCES

1. Politzer, I. R., Dowty, B. J., and Laseter, J. L., Use of gas chromatography and mass spectrometry to analyze underivatized volatile human or animal constituents of clinical interest, *Clin. Chem.*, 22, 1775, 1976.
2. Carroll, D. I., Dzidic, I., Stillwell, R. N., Haegele, K. D., and Horning, E. C., Atmospheric pressure ionization mass spectrometry: corona discharge ion source for use in liquid chromatograph-mass spectrometer-computer analytical system, *Anal. Chem.*, 47, 2369, 1975.
3. Baty, J. D. and Wade, A. P., Analysis of steroids in biological fluids by computer-aided gas-liquid chromatography-mass spectrometry, *Anal. Biochem.*, 57, 27, 1974.
4. Weinkam, R. J., Wen, J. H. C., Furst, D. E., and Levin, V. A., Analysis for 1,3-*bis*(2-chloroethyl)-1-nitrosourea by chemical ionization mass spectrometry, *Clin. Chem.*, 24, 45, 1978.
5. Lehrer, M. and Karmen, A., Quantitative analysis of diphenylhydantoin in serum by gas chromatography-mass spectrometry, *Biochem. Med.*, 14, 230, 1975.
6. Wilson, J. M., Williamson, L. J., and Raisys, V. A., Simultaneous measurement of secondary and tertiary tricyclic antidepressants by GC/MS chemical ionization mass spectrometry, *Clin. Chem.*, 23, 1012, 1977.
7. Dubois, J. P., Kung, W., Theobald, W., and Wirz, B., Measurement of clomipramine, N-desmethylclomipramine, imipramine, and dehydroimipramine in biological fluids by selective ion monitoring, and pharmacokinetics of clomipramine, *Clin. Chem.*, 22, 892, 1976.
8. Bjorkhem, I., Blomstrand, R., and Svensson, L., Serum cholesterol determination by mass fragmentography, *Clin. Chim. Acta*, 54, 185, 1974.
9. Frigerio, A. and Ghisalberti, E. L., Mass fragmentography (single and multiple ion monitoring) in drug research, in *Mass Spectrometry in Drug Metabolism,* Frigerio, A. and Ghisalberti, E. L., Eds., Plenum Press, New York, 1977, 131.
10. Gordon, A. E. and Frigerio, A., Mass fragmentography as an application of gas-liquid chromatography mass spectrometry in biological research, *J. Chromatogr.*, 73, 401, 1972.
11. Palmer, L., Bertilsson, L., Collst, P., and Rawlins, M., Quantitative determination of carbamazepine in plasma by mass fragmentography, *Clin. Pharm. Ther.*, 14, 827, 1973.
12. Zagalak, M. J., Curtius, H. C., Leimbacher, W., and Redweik, U., Quantitation of deuterated and nondeuterated phenylalanine and tyrosine in human plasma using the selective ion monitoring method with combined gas chromatography-mass spectrometry-application to the *in vivo* measurement of phenylalanine-4-monooxygenase activity, *J. Chromatogr.*, 142, 523, 1977.
13. Hachey, D. L., Kreek, M. J., and Mattson, D. H., Quantitative analysis of methadone in biological fluids using deuterium-labeled methadone and GLC-chemical-ionization mass spectrometry, *J. Pharm. Sci.*, 66, 1579, 1977.

14. Horning, M. G., Nowlin, J., Butler, C. M., Lertratanangkoon, K., Sommer, K., and Hill, R. M., Clinical applications of gas chromatography/mass spectrometer/computer systems, *Clin. Chem.*, 21, 1282, 1975.

15. Ghisalberti, E. L., Chemical ionization mass spectrometry in the identification of drugs and drug metabolites, in *Mass Spectrometry in Drug Metabolism*, Frigerio, A. and Ghisalberti, E. L., Eds., Plenum Press, New York, 1977, 291.

16. Murata, T. and Takahashi, S., Qualitative and quantitative chemical ionization mass spectrometry of triglycerides, *Anal. Chem.*, 49, 728, 1977.

17. Hatch, F. and Munson, B., Reactant ion monitoring for selective detection in gas chromatography/chemical ionization mass spectrometry, *Anal. Chem.*, 49, 731, 1977.

18. Jardine, I. and Fenselau, C., Charge exchange mass spectra of morphine and tropane alkaloids, *Anal. Chem.*, 47, 730, 1975.

19. Heath, H., *Drugs of Abuse*, Vol. 6, No. 2, U.S. Drug Enforcement Administration, Washington, D.C., 1980, 1.

20. Audier, H., Fetizon, M., Ginsburg, D., Mandelbaum, A., and Rull, T., Mass spectrometry of the morphine alkaloids, *Tetrahedron Lett.*, p. 13, 1965.

21. Mandelbaum, A. and Ginsburg, D., Studies in mass spectrometry. IV. Steric direction of fragmentation in *cis* and *trans*-B:C ring-fused morphine derivatives, *Tetrahedron Lett.*, p. 2479, 1965.

22. Smith, R. M., Forensic identification of opium by computerized gas chromatography/mass spectrometry, *J. Forensic Sci.*, 18, 327, 1973.

23. Cone, E. J., Darwin, W. D., and Buchwald, W. F., Assay for codeine, morphine and ten potential urinary metabolites by gas chromatography-mass fragmentography, *J. Chromatogr.*, 275, 307, 1983.

24. Cone, E. J., Gorodetzky, C. W., Yeh, S. Y., Darwin, W. D., and Buchwald, W. F., Detection and measurement of opium alkaloids and metabolites in urine of opium eaters by methane chemical ionization mass fragmentography, *J. Chromatogr.*, 230, 57, 1982.

25. Dahlstrom, B. and Paalzow, L., Quantitative determination of morphine in biological samples by gas-liquid chromatography and electron-capture detection, *J. Pharm. Pharmacol.*, 27, 172, 1975.

26. Felby, S., Morphine: its quantitative determination in nanogram amounts in small samples of whole blood by electron-capture gas chromatography, *Forensic Sci. Int.*, 13, 145, 1979.

27. Garrett, E. R. and Gurkan, T., Pharmacokinetics of morphine and its surrogates. I. Comparisons of sensitive assays of morphine in biological fluids and application to morphine pharmacokinetics in the dog, *J. Pharm. Sci.*, 67, 1512, 1978.

28. Plomp, G. J. J., Maes, R. A. A., and Van Ree, J., Disposition of morphine in rat brain: relationship to biological activity, *J. Pharmacol. Exp. Ther.*, 217, 181, 1981.

29. Wallace, J. E., Hamilton, H. E., Blum, K., and Petty, C., Determination of morphine in biological fluids by electron capture gas-liquid chromatography, *Anal. Chem.*, 46, 2107, 1974.

30. Jones, A. W., Blom, Y., and Bondesson, U., Determination of morphine in biological samples by gas chromatography-mass spectrometry. Evidence for persistent tissue binding in rats twenty-two days post-withdrawal, *J. Chromatogr.*, 309, 73, 1984.

31. Drost, R. H., Van Ooijen, R. D., Ionescu, T., and Maes, R. A. A., Determination of morphine in serum and cerebrospinal fluid by gas chromatography and selected ion monitoring after reversed-phase column extraction, *J. Chromatogr.*, 310, 193, 1984.

32. Cone, E. J., Darwin, W. D., Buchwald, W. F., and Gorodetzky, C. W., Oxymorphone metabolism and urinary excretion in human, rat, guinea pig, rabbit, and dog, *Drug Metab. Dispos.*, 11, 446, 1983.

33. Jindal, S. P. and Vestergaard, P., Quantitation of etorphine in urine by selected ion monitoring using tritiated etorphine as an internal standard, *J. Pharm. Sci.*, 67, 260, 1978.

34. Ebbighausen, W. O. R., Mowat, J. H., Stearns, H., and Vestergaard, P., Mass fragmentography of morphine and 6-monoacetylmorphine in blood with a stable isotope internal standard, *Biomed. Mass Spectrom.*, 1, 305, 1974.

35. Clarke, P. A. and Foltz, R. L., Quantitative analysis of morphine in urine by gas chromatography-chemical ionization-mass spectrometry, with [N-C^{2}H$_3$] morphine as an internal standard, *Clin. Chem.*, 20, 465, 1974.

36. Cone, E. J., Gorodetzky, C. W., Yousefnejad, D., Buchwald, W. F., and Johnson, R. E., The metabolism and excretion of buprenorphine in humans, *Drug Metab. Dispos.*, 12, 577, 1984.

37. Blom, Y., Bondesson, U., and Anggard, E., Analysis of buprenorphine and its N-dealkylated metabolite in plasma and urine by selected-ion monitoring, *J. Chromatogr.*, 338, 89, 1985.

38. Evans, J. V., Leeling, J. L., and Helms, R. J., Methodology for the identification of the urinary metabolites of the analgesic-narcotic antagonist, codorphone, by gas chromatography mass spectrometry, *Biomed. Mass Spectrom.*, 9, 191, 1982.

39. Goromaru, T., Matsuura, H., Furuta, T., Baba, S., Yoshimura, N., Miyawaki, T., and Sameshima, T., Identification of fentanyl metabolites in rat urine by gas chromatography-mass spectrometry with stable-isotope tracers, *Drug Metab. Dispos.*, 10, 542, 1982.

40. Lin, S. N., Wang, T. F., Caprioli, R. M., and Mo, B. P. N., Determination of plasma fentanyl by gc-mass spectrometry and pharmacokinetic analysis, *J. Pharm. Sci.,* 70, 1276, 1981.
41. Van Rooy, H. H., Vermeulen, N. P. E., and Bovill, J. G., The assay of fentanyl and its metabolites in plasma of patients using gas chromatography with alkali flame ionization detection and gas chromatography-mass spectrometry, *J. Chromatogr.,* 223, 85, 1981.
42. Garriott, J. C., Rodriguez, R., and DiMaio, V. J. M., A death from fentanyl overdose, *J. Anal. Toxicol.,* 8, 288, 1984.
43. Min, B. H., Garland, W. A., and Pao, J., Determination of levorphanol (Levo-Dromoran®) in human plasma by combined gas chromatography-negative ion chemical ionization mass spectrometry, *J. Chromatogr.,* 231, 194, 1982.
44. Cole, W. J., Parkhouse, J., and Yousef, Y. Y., Application of the extractive alkylation technique to the pentafluorobenzylation of morphine (a heroin metabolite) and surrogates, with special reference to the quantitative determination of plasma morphine levels using mass fragmentography, *J. Chromatogr.,* 136, 409, 1977.
45. Herman, R. J., McAllister, C. B., Branch, R. A., and Wilkinson, G. R., Effects of age on meperidine disposition, *Clin. Pharmacol. Ther.,* 37, 19, 1985.
46. Verbeeck, R. K., James, R. C., Taber, D. F., Sweetman, B. J., and Wilkinson, G. R., The determination of meperidine, normeperidine and deuterated analogs in blood and plasma by gas chromatography mass spectrometry selected ion monitoring, *Biomed. Mass Spectrom.,* 7, 58, 1980.
47. Yeh, S. Y., Metabolism of meperidine in several animal species, *J. Pharm. Sci.,* 73, 1783, 1984.
48. Bondesson, U. and Hartvig, P., Mass fragmentographic method for the determination of ketobemidone in plasma, *J. Chromatogr.,* 179, 207, 1979.
49. Bondesson, U., Hartvig, P., Abrahamsson, L., and Ahnfelt, N. D., Simultaneous determination of ketobemidone and its N-demethylated metabolite in patient plasma samples by gas chromatography mass spectrometry with selected ion monitoring, *Biomed. Mass Spectrom.,* 10, 283, 1983.
50. Bondesson, U., Hartvig, P., and Danielsson, B., Quantitative determination of the urinary excretion of ketobemidone and four of its metabolites after intravenous and oral administration in man, *Drug Metab. Dispos.,* 9, 376, 1981.
51. Bellward, G. D., Warren, P. M., Howald, W., Axelson, J. E., and Abbott, F. S., Methadone maintenance: effect of urinary pH on renal clearance in chronic high and low doses, *Clin. Pharmacol. Ther.,* 22, 92, 1977.
52. Kang, G. I. and Abbott, F. S., Analysis of methadone and metabolites in biological fluids with gas chromatography-mass spectrometry, *J. Chromatogr.,* 231, 311, 1982.
53. Sullivan, H. R., Marshall, F. J., McMahon, R. E., Anggard, E., Gunne, L. M., and Holmstrand, J. H., Mass fragmentographic determination of unlabeled and deuterium-labeled methadone in human plasma — possibilities for measurement of steady state pharmacokinetics, *Biomed. Mass Spectrom.,* 2, 197, 1975.
54. Rust, M., Improper marijuana testing charged, *Am. Med. News,* 1, 1984.
55. Mason, A. P., Perez-Reyes, M., McBay, A. J., and Foltz, R. L., Cannabinoid concentrations in plasma after passive inhalation of marijuana smoke, *J. Anal. Toxicol.,* 7, 172, 1983.
56. Wethe, G., Bugge, A., Bones, T., Morland, J., Skuterud, A., and Steen, A., Passive smoking of cannabis, *Acta Pharm. Toxicol.,* 51, 21, 1982.
57. Harvey, D. J. and Paton, W. D. M., Use of trimethylsilyl and other homologous trialkylsilyl derivatives for the separation and characterization of mono- and dihydroxy-cannabinoids by combined gas chromatography and mass spectrometry, *J. Chromatogr.,* 109, 73, 1975.
58. Harvey, D. J., Cyclic alkylboronates as derivatives for the characterization of cannabinolic acids by combined gas chromatography and mass spectrometry, *Biomed. Mass Spectrom.,* 4, 88, 1977.
59. Quilliam, M. A. and Westmore, J. B., Sterically crowded trialkylsilyl derivatives for chromatography and mass spectrometry of biologically important compounds, *Anal. Chem.,* 50, 59, 1978.
60. Agurell, S., Gustafsson, B., Holmstedt, B., Leander, K., Lindgren, J. E., Nilsson, I., Sandberg, F., and Asberg, M., Quantitation of Δ^1-tetrahydrocannabinol in plasma from cannabis smokers, *J. Pharm. Pharmacol.,* 25, 554, 1973.
61. Rosenthal, D., Harvey, T. M., Bursey, J. T., Brine, D. R., and Wall, M. E., Comparison of gas chromatography mass spectrometry methods for the determination of Δ^9-tetrahydrocannabinol in plasma, *Biomed. Mass Spectrom.,* 5, 312, 1978.
62. Pirl, J. N., Papa, V. M., and Spikes, J. J., The detection of *delta-9*-tetrahydrocannabinol in postmortem blood samples, *J. Anal. Toxicol.,* 3, 129, 1979.
63. Harvey, D. J., Leuschner, J. T. A., and Paton, W. D. M., Measurement of Δ^1-tetrahydrocannabinol in plasma to the low picogram range by gas chromatography-mass spectrometry using metastable ion detection, *J. Chromatogr.,* 202, 83, 1980.

64. Jones, A. B., El Sohly, H. N., Arafat, E. S., and El Sohly, M. A., Analysis of the major metabolites of Δ⁹-tetrahydrocannabinol in urine. IV. A comparison of five methods, *J. Anal. Toxicol.*, 8, 249, 1984.

65. Irving, J., Leeb, B., Foltz, R. L., Cook, C. E., Bursey, J. T., and Willette, R. E., Evaluation of immunoassays for cannabinoids in urine, *J. Anal. Toxicol.*, 8, 192, 1984.

66. Ohlsson, A., Lindgren, J. E., Wahlen, A., Agurell, S., Hollister, L. E., and Gillespie, H. K., Single dose kinetics of deuterium labelled Δ¹-tetrahydrocannabinol in heavy and light cannabis users, *Biomed. Mass Spectrom.*, 9, 6, 1982.

67. Baker, T. S., Harry, J. V., Russell, J. W., and Myers, R. L., Rapid method for the gc/ms confirmation of 11-nor-9-carboxy-Δ⁹-tetrahydrocannabinol in urine, *J. Anal. Toxicol.*, 8, 255, 1984.

68. Foltz, R. L., McGinnis, K. M., and Chinn, D. M., Quantitative measurement of Δ⁹-tetrahydrocannabinol and two major metabolites in physiological specimens using capillary column gas chromatography negative ion chemical ionization mass spectrometry, *Biomed. Mass Spectrom.*, 10, 316, 1983.

69. Green, D. E., Quantitation of cannabinoids in biological specimens using probability based matching gas chromatography/mass spectrometry, in *Cannabinoid Assays in Humans NIDA Research Monograph 7*, Willette, R. E., Ed., National Institute on Drug Abuse, Rockville, Md., 1976, 70.

70. Nordqvist, M., Lindgren, J. E., and Agurell, S., A method for the identification of acid metabolites of tetrahydrocannabinol (THC) by mass fragmentography, in *Cannabinoid Assays in Humans NIDA Research Monograph 7*, Willette, R. E., Ed., National Institute on Drug Abuse, Rockville, Md., 1976, 64.

71. Harvey, D. J., Leuschner, J. T. A., and Paton, W. D. M., Gas chromatographic and mass spectrometric studies on the metabolism and pharmacokinetics of Δ¹-tetrahydrocannabinol in the rabbit, *J. Chromatogr.*, 239, 243, 1982.

72. Van Langenhove, A., Biller, J. E., Biemann, K., and Browne, T. R., Simultaneous determination of phenobarbital and *p*-hydroxyphenobarbital and their stable isotope labeled analogs by gas chromatography mass spectrometry, *Biomed. Mass Spectrom.*, 9, 201, 1982.

73. Dykeman, R. W. and Ecobichon, D. J., Simultaneous determination of phenytoin, phenobarbital and their para-hydroxylated metabolites in urine by reversed-phase high-performance liquid chromatography, *J. Chromatogr.*, 162, 104, 1979.

74. Hooper, W. D., Kunze, H. E., and Eadie, M. J., Simultaneous assay of methylphenobarbital and phenobarbital in plasma using gas chromatography-mass spectrometry with selected ion monitoring, *J. Chromatogr.*, 223, 426, 1981.

75. Kallberg, N., Agurell, S., Ericsson, O., Bucht, E., Jalling, B., and Boreus, L. O., Quantitation of phenobarbital and its main metabolites in human urine, *Eur. J. Clin. Pharmacol.*, 9, 311, 1982.

76. Whyte, M. P. and Dekaban, A. D., Metabolic fate of phenobarbital — a quantitative study of *p*-hydroxyphenobarbital elimination in man, *Drug Metab. Dispos.*, 5, 63, 1977.

77. Tang, B. K., Yilmaz, B., and Kalow, W., Determination of phenobarbital, *p*-hydroxyphenobarbital and phenobarbital-N-glucoside in urine by gas chromatography chemical ionization mass spectrometry, *Biomed. Mass Spectrom.*, 11, 462, 1984.

78. Truscott, R. J. W., Burke, D. G., Korth, J., Halpern, B., and Summons, R., Simultaneous determination of diphenylhydantoin, mephobarbital, carbamazepine, phenobarbital and primidone in serum using direct chemical ionization mass spectrometry, *Biomed. Mass Spectrom.*, 5, 477, 1978.

79. Kupferberg, H. J. and Longacre-Shaw, J., Mephobarbital and phenobarbital plasma concentrations in epileptic patients treated with mephobarbital, *Ther. Drug. Monit.*, 1, 117, 1979.

80. Heusler, H., Epping, J., Heusler, S., and Richter, E., Simultaneous determination of blood concentrations of methohexital and its hydroxy metabolite by gas chromatography and identification of 4-hydroxymethohexital by combined gas-liquid chromatography-mass spectrometry, *J. Chromatogr.*, 226, 403, 1981.

81. Aderjan, V. R., Fritz, P., and Mattern, R., Zur Bedeutung des Nachweises und der Pharmakokinetik von Flurazepam-Metaboliten in menschlichem Blut, *Arzneim. Forsch.*, 30, 1944, 1980.

82. Schwartz, M. A. and Postma, E., Metabolism of flurazepam, a benzodiazepine, in man and dog, *J. Pharm. Sci.*, 59, 1800, 1970.

83. Schwartz, M. A., Vane, F. M., and Postma, E., Urinary metabolites of 7-chloro-1-(2-diethylaminoethyl)-5-(2-fluorophenyl)-1,3-dihydro-2H-1,4-benzodiazepin-2-one dihydrochloride, *J. Med. Chem.*, 11, 770, 1968.

84. Garland, W. A. and Miwa, B. J., The [M − 1]⁻ ion in the negative chemical ionization mass spectra of diazepam and nordiazepam, *Biomed. Mass Spectrom.*, 10, 126, 1983.

85. Garland, W. A., Miwa, B. J., Diarman, W., Kappell, B., Chiueh, M. C. C., Divoll, M., and Greenblatt, D. J., Identification of 7-chloro-5-(2'-fluorophenyl)-2,3-dihydro-2-oxo-1H-1,4-benzodiazepine-1-acetaldehyde, a new metabolite of flurazepam in man, *Drug Metab. Dispos.*, 11, 70, 1983.

86. Vasiliades, J. and Sahawneh, T., Midazolam determination by gas chromatography, liquid chromatography and gas chromatography-mass spectrometry, *J. Chromatogr.*, 228, 195, 1982.

87. Rubio, F., Miwa, B. J., and Garland, W. A., Determination of midazolam and two metabolites of midazolam in human plasma by gas chromatography-negative chemical-ionization mass spectrometry, *J. Chromatogr.*, 233, 157, 1982.

88. Min, B. H. and Garland, W. A., Determination of clonazepam and its 7-amino metabolite in plasma and blood by gas chromatography-chemical ionization mass spectrometry, *J. Chromatogr.*, 139, 121, 1977.

89. Brettel, T. A., Analysis of N-mono-trifluoroacetyl derivatives of amphetamine analogues by gas chromatography and mass spectrometry, *J. Chromatogr.*, 257, 45, 1983.

90. Foltz, R. L., Fentiman, A. F., and Foltz, R. B., GC/MS assays for abused drugs in body fluids, *NIDA Research Monograph Number 32*, National Institute on Drug Abuse, Rockville, Md., 1980.

91. Kojima, T., Yashiki, M., Une, I., Noda, J., Tsukue, I., and Sakai, K., Articles found in the possession of a methamphetamine abuser, *Forensic Sci. Int.*, 26, 207, 1984.

92. Kojima, T., Une, I., and Yashiki, M., CI-mass fragmentographic analysis of methamphetamine and amphetamine in human autopsy tissues after acute methamphetamine poisoning, *Forensic Sci. Int.*, 21, 253, 1983.

93. Reynolds, G. P. and Gray, D. O., Gas chromatographic detection of N-methyl-2-phenylethylamine: a new component of human urine, *J. Chromatogr.*, 145, 137, 1978.

94. Cimbura, G. and Koefoed, J., A review of some glc-fid derivatization techniques found useful in forensic toxicology, *J. Chromatogr. Sci.*, 12, 261, 1974.

95. Hiemke, C., Kauert, G., and Kalbhen, D. A., Gas-liquid chromatographic properties of catecholamines, phenylethylamines, and indolalkylamines as their propionyl derivatives, *J. Chromatogr.*, 153, 451, 1978.

96. Narasimhachari, N., Friedel, R. O., Schlemmer, F., and Davis, J. M., Quantitation of amphetamine in plasma and cerebrospinal fluid by gas chromatography-mass spectrometry selected ion monitoring, using β-methylphenethylamine as an internal standard, *J. Chromatogr.*, 164, 386, 1979.

97. Cho, A. H., Lindeke, B., Hedhson, B. J., and Jenden, D. J., Deuterium substituted amphetamine as an internal standard in a gas chromatographic/mass spectrometric (gc/ms) assay for amphetamine, *Anal. Chem.*, 45, 570, 1973.

98. Coutts, R. T., Dawe, R., Jones, G. R., Liu, S. F., and Midha, K. N., Analysis of perfluoroacyl derivatives of ephedrine, pseudoephedrine and analogues by gas chromatography and mass spectrometry, *J. Chromatogr.*, 190, 53, 1980.

99. Coutts, R. T., Baker, G. B., Pasutto, F. M., Liu, S. F., LeGatt, D. F., and Prelusky, D. B., Mass spectral analysis of perfluoroacylated derivatives of some arylalkylamines of biological interest, *Biomed. Mass Spectrom.*, 11, 441, 1984.

100. Delbeke, F. T., DeBackere, M., Jonckheere, J. A. A., and DeLeenheer, A. P., Pentafluorobenzoyl derivatives of doping agents. I. Extractive benzoylation and gas chromatography with electron capture detection of primary and secondary amines, *J. Chromatogr.*, 273, 141, 1983.

101. Tas, A. C., Odink, J., Ten Noever de Brauw, M. C., Schrijver, J., and Jonk, R. J. G., Derivatization and mass spectrometric behavior of catecholamines and their 3-O-methylated metabolites, *J. Chromatogr.*, 310, 243, 1984.

102. Ehrhardt, J. D. and Schwarz, J., A gas chromatography-mass spectrometry assay of human plasma catecholamines, *Clin. Chim. Acta*, 88, 71, 1978.

103. Ohki, T., Saito, A., Ohta, K., Niwa, T., Maeda, K., and Sakakibara, J., Amine metabolite profile of normal and uremic urine using gas chromatography-mass spectrometry, *J. Chromatogr.*, 233, 1, 1982.

104. Hsu, L. S. F., Sharrard, J. I., Love, C., and Marrs, T. C., A rapid method for screening urine samples in suspected abuse of cocaine, *Ann. Clin. Biochem.*, 18, 368, 1981.

105. Ambre, J. J., Ruo, T. I., Smith, G. L., Backes, D., and Smith, C. M., Ecgonine methyl ester, a major metabolite of cocaine, *J. Anal. Toxicol.*, 6, 26, 1982.

106. Chinn, D. M., Crouch, D. J., Peat, M. A., Finkle, B. S., and Jennison, T. A., Gas chromatography-chemical ionization mass spectrometry of cocaine and its metabolites in biological fluids, *J. Anal. Toxicol.*, 4, 37, 1980.

107. Graffeo, A. P., Lin, D. C. K., and Foltz, R. L., Analysis of benzoylecgonine in urine by high-performance liquid chromatography and gas chromatography-mass spectrometry, *J. Chromatogr.*, 126, 717, 1976.

108. Jindal, S. P. and Verstergaard, P., Quantitation of cocaine and its principal metabolite, benzoylecgonine, by glc-mass spectrometry using stable isotope labeled analogs as internal standards, *J. Pharm. Sci.*, 67, 811, 1978.

109. Matsubara, K., Kagawa, M., and Fukui, Y., In vivo and in vitro studies on cocaine metabolism: ecgonine methyl ester as a major metabolite of cocaine, *Forensic Sci. Int.*, 26, 169, 1984.

110. Matsubara, K., Maseda, C., and Fukui, Y., Quantitation of cocaine, benzoylecgonine and ecgonine methyl ester by gc-ci-sim after Extrelut® extraction, *Forensic Sci. Int.*, 26, 181, 1984.

111. Pearce, D. S., Detection and quantitation of phencyclidine in blood by use of [²H₅]phencyclidine and selected ion monitoring applied to non-fatal cases of phencyclidine intoxication, *Clin. Chem.*, 22, 1623, 1976.

112. Baker, J. K. and Little, T. L., Metabolism of phencyclidine. The role of the carbinolamine intermediate in the formation of lactam and amino acid metabolites of nitrogen heterocycles, *J. Med. Chem.*, 28, 46, 1985.

113. Wong, L. K. and Biemann, K., Metabolites of phencyclidine, *Clin. Toxicol.*, 9, 583, 1976.

114. Kuhnert, B. R., Bagby, B. S., and Golden, N. L., Measurement of phencyclidine and two hydroxylated metabolites by selected ion monitoring, *J. Chromatogr.*, 276, 433, 1983.

115. Domino, E. F., Neurobiology of phencyclidine — an update, in *Phencyclidine (PCP) Abuse: An Appraisal, NIDA Research Monograph Number 21*, Petersen, R. C. and Stillman, R. C., Eds., National Institute on Drug Abuse, Rockville, Md., 1978, 18.

116. Ward, D. P., Trevor, A. J., Kalir, A., Adams, J. D., Baillie, T. A., and Castagnoli, N., Jr., Metabolism of phencyclidine — the role of iminium ion formation in covalent binding to rabbit microsomal protein, *Drug Metab. Dispos.*, 10, 690, 1982.

117. Cook, C. E., Brine, D. R., and Tallent, C. R., Identification of *in vitro* rat metabolites of 1-phenylcyclohexene, *Drug Metab. Dispos.*, 12, 186, 1984.

118. Callaham, M. and Kassell, D., Epidemiology of fatal tricyclic antidepressant ingestion: implications for management, *Ann. Emerg. Med.*, 14, 29, 1985.

119. Belvedere, G., Burti, L., Frigerio, A., and Pantarotto, C., Gas chromatographic-mass fragmentographic determination of "steady-state" plasma levels of imipramine and desipramine in chronically treated patients, *J. Chromatogr.*, 111, 313, 1975.

120. Frigerio, A., Belvedere, G., DeNadai, F., Fanelli, R., Pantarotto, C., Riva, E., and Morselli, P. L., A method for the determination of imipramine in human plasma by gas-liquid chromatography-mass fragmentography, *J. Chromatogr.*, 74, 201, 1972.

121. Ishida, R., Ozaki, T., Uchida, H., and Irikura, T., Gas chromatographic-mass spectrometric determination of amitriptyline and its major metabolites in human serum, *J. Chromatogr.*, 305, 73, 1984.

122. Garland, W. A., Muccino, R. R., Min, B. H., Cupano, J., and Fann, W. E., A method for the determination of amitriptyline and its metabolites, nortriptyline, 10-hydroxyamitriptyline and 10-hydroxynortriptyline in human plasma using stable isotope dilution and gas chromatography-chemical ionization mass spectrometry (gc-cims), *Clin. Pharmacol. Ther.*, 25, 844, 1979.

123. Craig, J. C., Gruenke, L. D., and Nguyen, T. L., Simultaneous analysis of imipramine and its metabolite desipramine in biological fluids, *J. Chromatgr.*, 239, 81, 1982.

124. Davis, T. P., Veggeberg, S. K., Hameroff, S. R., and Watts, K. L., Sensitive and quantitative determination of plasma doxepin and desmethyldoxepin in chronic pain patients by gas chromatography and mass spectrometry, *J. Chromatogr.*, 273, 436, 1983.

125. Tasset, J. J. and Pesce, A. J., Amoxapine in human overdose, *J. Anal. Toxicol.*, 8, 124, 1984.

126. Maurer, H. and Pfleger, K., Determination of 1,4- and 1,5-benzodiazepines in urine using a computerized gas chromatographic-mass spectrometric technique, *J. Chromatogr.*, 222, 409, 1981.

127. Maurer, H. and Pfleger, K., Screening procedure for detection of antidepressants and their metabolites in urine using a computerized gas chromatographic-mass spectrometric technique, *J. Chromatogr.*, 305, 309, 1984.

128. Maurer, H. and Pfleger, K., Screening procedure for detection of phenothiazine and analogous neuroleptics and their metabolites in urine using a computerized gas chromatographic-mass spectrometric technique, *J. Chromatogr.*, 306, 124, 1984.

129. Maurer, H. and Pfleger, K., Verfahren zum Nachweis von Psychopharmaka und Suchtstoffen im Urin mit einer Computer-unterstutzten Gas-chromatographie-Massenspektometrie-Technik (gc-msds), *J. Clin. Chem. Clin. Biochem.*, 20, 667, 1982.

130. Chan, K. Y. and Okerholm, R. A., Quantitative analysis of melperone in human plasma by gas chromatography-mass spectrometry-selected ion monitoring, *J. Chromatogr.*, 274, 121, 1983.

131. Falkner, F. C., Fouda, H. G., and Mullins, F. G., A gas chromatographic/mass spectrometric assay for flutroline, a γ-carboline antipsychotic agent, with direct derivatization on a moving needle injector, *Biomed. Mass Spectrom.*, 11, 482, 1984.

132. Semple, D. J., Zimelidine distribution in a sudden death, *J. Anal. Toxicol.*, 8, 285, 1984.

133. Stajic, M., Granger, R. H., and Beyer, J. C., Fatal metoprolol overdose, *J. Anal. Toxicol.*, 8, 228, 1984.

134. Ervik, M., Hoffmann, K. J., and Kylberg-Hanssen, K., Selected ion monitoring of metoprolol and two metabolites in plasma and urine using deuterated internal standards, *Biomed. Mass Spectrom.*, 8, 322, 1981.

135. Anderson, P., Noher, H., and Swahn, C. G., Pharmacokinetics of baclofen overdose, *Clin. Toxicol.*, 22, 11, 1984.

136. Swahn, C. G., Beving, H., and Sedvall, G., Mass fragmentographic determination of 4-amino-3-*p*-chlorophenylbutyric acid (baclofen) in cerebrospinal fluid and serum, *J. Chromatogr.*, 162, 433, 1979.

137. Degen, P. H. and Riess, W., The determination of γ-amino-β-(*p*-chlorophenyl)butyric acid (baclofen) in biological material by gas-liquid chromatography, *J. Chromatogr.*, 117, 399, 1976.

138. Holzbecher, M., Perry, R. A., and Ellenberger, H. A., Report of a metoprolol-associated death, *J. Forensic Sci.*, 27, 715, 1982.

139. Möller, B. H. J., Massive intoxication with metroprolol, *Br. Med. J.*, 1, 222, 1976.

140. Shore, E. T., Cepin, D., and Davidson, M. J., Metoprolol overdose, *Ann. Emerg. Med.*, 10, 524, 1981.

141. Sire, S., Metoprolol intoxication, *Lancet*, 2, 1137, 1973.

142. Bridges, R. R. and Jennison, T. A., Analysis of ethchlorvynol (Placidyl®): evaluation of a comparison performed in a clinical laboratory, *J. Anal. Toxicol.*, 8, 263, 1984.

143. Horwitz, J. P., Brukwinski, W., Treisman, J., Andrzejewski, D., Hills, E. B., Chung, H. L., and Wang, C. Y., Ethchlorvynol: potential of metabolites for adverse effects in man, *Drug Metab. Dispos.*, 8, 77, 1980.

144. Yost, R. A., Fetterolf, D. D., Hass, J. R., Harvan, D. J., Weston, A. F., Skotnicki, P. A., and Simon, N. M., Comparison of mass spectrometric methods for trace level screening of hexachlorobenzene and trichlorophenol in human blood serum and urine, *Anal. Chem.*, 56, 2223, 1984.

145. Brotherton, H. O. and Yost, R. A., Determination of drugs in blood serum by mass spectrometry/mass spectrometry, *Anal. Chem.*, 55, 549, 1983.

146. Wright, L. H., Edgerton, T. R., Arbes, S. J., Jr., and Lores, E. M., The determination of underivatized chlorophenols in human urine by combined high performance liquid chromatography mass spectrometry and selected ion monitoring, *Biomed. Mass Spectrom.*, 8, 475, 1981.

147. Robert, T. A. and Hagardorn, A. N., Analysis and kinetics of 2,4-dinitrophenol in tissues by capillary gas chromatography-mass spectrometry, *J. Chromatogr.*, 276, 77, 1983.

148. Tanaka, A., Nose, N., Masaki, H., and Iwasaki, H., Determination of the stable isotope of nitrite flux in the blood of mice by gas chromatography-mass spectrometry with selected ion monitoring, *J. Chromatogr.*, 306, 51, 1984.

149. Anderson, M. M., Mays, J. B., Mitchum, R. K., and Hinson, J. A., Metabolism of 4-nitroaniline by rat liver microsomes, *Drug Metab. Dispos.*, 12, 179, 1984.

150. Lau, S. S., Monks, T. J., and Gillette, J. R., Identification of 2-bromohydroquinone as a metabolite of bromobenzene and *o*-bromophenol: implications for bromobenzene-induced nephrotoxicity, *J. Pharmacol. Exp. Ther.*, 230, 360, 1984.

151. Monks, T. J., Lau, S. S., Pohl, L. R., and Gillette, J. R., The mechanism of formation of *o*-bromophenol from bromobenzene, *Drug Metab. Dispos.*, 12, 193, 1984.

152. Kropscott, B. E., Kastl, P. E., and Herman, E. A., Determination of 2-ethylhexyl 2,4-dichlorophenoxyacetate in rat blood and urine by electron-capture gas chromatography and gas chromatography-mass spectrometry, *J. Chromatogr.*, 203, 1984.

153. Bardalaye, P. C., Wheeler, W. B., and Templeton, J. L., Derivatization of Oryzalin for characterization by gas chromatography-mass spectrometry, *J. Chromatogr.*, 314, 450, 1984.

154. Cairns, T., Siegmund, E. G., and Bong, R. L., Chemical ionization mass spectrometry of isofenphos and its metabolites, *Anal. Chem.*, 56, 2547, 1984.

155. Singh, A. K., Improved analysis of acephate and methamidophos in biological samples by selective ion monitoring gas chromatography-mass spectrometry, *J. Chromatogr.*, 301, 465, 1984.

Chapter 3

MASS SPECTROMETRY IN SPORTS TESTING

Jack Henion, Dominique Silvestre, and George Maylin

TABLE OF CONTENTS

I. INTRODUCTION

Public interest in competition sports events has never been higher than it is today. The range of events as demonstrated in the recent Los Angeles Olympic Games varies from Roman-Greco wrestling to equestrian competition. In spite of seemingly unbeatable established records such as marathon times and pole vaulting heights, new methods of training and extraordinary personal efforts continually slash the old records and establish new goals to conquer. It is certainly the thrill of witnessing the "impossible" achievement or the cheering of a favorite athlete or team to victory that draws millions of spectators to sports events throughout the world.

Although so-called amateur athletes compete without compensation and train with little financial support, the tremendous publicity and concomitant commercialization of even amateur competition events has produced considerable pressure for the "need" to win. When this is coupled with the natural tendency for mankind to wager on the outcome of a sports event, we have an environment which tends to cause the competition to resort to "winning schemes" beyond the classical or even modern training regimes. These schemes increasingly include the administration of drugs to the athlete prior to or during the sports event which may be considered helpful to the individual during competition.

The use of drugs outside the care or guidance of a physician is considered a growing problem in our schools and the workplace. However, the apparent common use of certain drugs in amateur and professional sports events has elicited concern by those groups or individuals who are concerned with the integrity of sports. Since the use of certain drugs may provide some competitive advantage over the comparable athlete who does not use drugs, there is concern with the potential for unfair competition. There is also concern for the health and welfare of athletes who use drugs for competitive advantage. The complications become extensive when the winner who secretly used drugs receives a $100,000 purse or worse yet dies of cardiac arrest during the sports event because of an adverse drug reaction.[1]

The use of drugs by sports contestants who wish to gain a competitive advantage has resulted in increased interest in implementing drug testing procedures. Although drug testing has not been extensively implemented in human events in the U.S., they have been utilized in recent Olympic Games and the Pan American Games held during the summer of 1983, as well as other sporting events.

In contrast, drug testing procedures have been in place for many years in animal competitions such as horse and dog racing. In fact, routine racehorse drug testing procedures using modern analytical technologies are now in place throughout the U.S., Canada, Europe, China, Australia, and other parts of the world today.

Horseracing has historically been considered the sport of kings and the integrity of racing has always been of paramount importance. The racing industry has sanctioned the implementation of racehorse drug testing to protect the human participant, the horse, and the outcome of the race. It is important to note, too, that the racing industry has provided generous financial support for drug testing research and development. It has been this commitment that has provided the techniques and experience available today for broad-scale drug testing in the competition sports arena.

One important difference between testing for drugs in horses, for example, vs. humans is that the legal ramifications of the individual's rights are not a factor when urine or blood are collected from a horse. This seemingly subtle difference has significant legal and forensic ramifications for the drug testing and sports committee personnel. Simply put, if the particular racing jurisdiction agrees to utilize drug testing procedures it becomes a rule by which all participants must abide. The horse is not required to give consent to the collection of a urine sample and it does not argue about invasion of its privacy as a human sports participant might do.

Aside from this, it is vitally important that all aspects of a drug testing and confirmation protocol be conducted in a professional, highly competent manner. The report of a "positive" finding can have a significant negative impact upon the personnel involved. In addition to personal embarrassment, a positive drug finding can disqualify the participant from future events, cause the loss of the honors achieved in the event, and result in substantial financial losses. These unfortunate occurrences, in addition to the unfavorable publicity to the sports industry itself, make positive drug findings a matter of serious concern.

We have chosen mainly racehorse drug testing with illustrative examples of modern analytical chemistry and, in particular, mass spectrometry (MS) to demonstrate the determination of foreign substances in the blood and urine of racehorses. Because there is often considerable money and professional reputation at stake, all aspects of the "test" from sample collection to analytical confirmation must be done with extreme care and professionalism. The identity and integrity of the samples must be maintained at all times or the analytical finding may be deemed useless.

The laboratory and person responsible for reporting a positive analytical finding in a toxicological sample is faced with an important scientific and legal task. Scientifically, the methodology and equipment must be capable of enduring legal scrutiny more stringent than peer review typical of refereed journal manuscripts. In addition, the chain of evidence from sample collection to final confirmation must be intact and completely documented. All that is needed to negate valid analytical results in a forensic case could be demonstration of a possible error in the scientific methodology or inaccurate results from the analytical instrumentation used to confirm the identity of an illicit chemical compound.

The care and detail required for reporting a new chemical structure in the scientific literature is not far removed from that required in a forensic case. However, the former is unlikely to be challenged in a court of law. For a new chemical compound to be reported, laborious successive crystallization for X-ray crystallographic measurements, spectroscopic measurements including IR, NMR, and MS, chromatographic separation, combustion analysis, or high-resolution MS measurements, plus a battery of chemical reaction sequences, are required to demonstrate the chemical behavior of the proposed structure. The same degree of scientific scrutiny should be applied in a forensic analysis. However, many of the chemical techniques used for structural elucidation are not useful for forensic analysis because of cost, time, and sensitivity limitations.

The analytical problems are particularly difficult today in organic trace analysis because the analytes of interest often are present at very low levels in a complex matrix. At the present time the most definitive analytical technique for such problems is MS. The mass spectrometer offers very high sensitivity and specificity for organic compounds.[2] When the mass spectrometer is coupled with an on-line chromatographic technique such as gas chromatography (GC) or high-pressure liquid chromatography (HPLC), the combination is very powerful and useful.[2,3]

II. EXPERIMENTAL PROCEDURE

For any forensic analysis the methodology and equipment performance should be adequately evaluated and documented. Quality control (QC) refers to those actions taken in the laboratory to insure that the measurement system is operating properly.[4] Examples include analyzing reference standards through the system, calibrating the instrument, keeping quality control charts, etc. Quality assurance (QA) refers to the

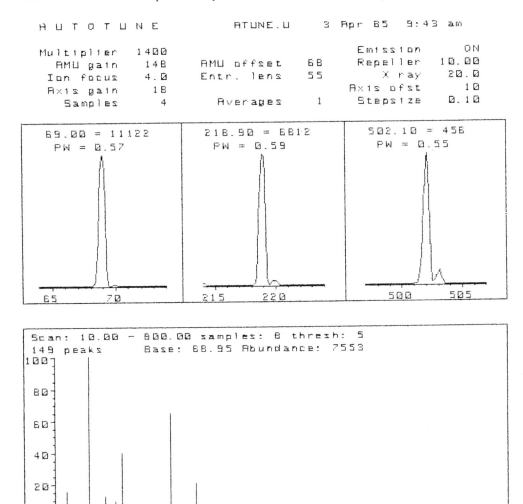

FIGURE 1. Tuning parameters, ion peak shapes, and plotted mass spectrum of calibration compound, perfluorotributylamine (PFTBA).

program whereby management assures itself (and its clients) that the quality control measures are being applied and that the results do in fact refer to the sample that has been submitted or collected for the laboratory.[4a,b]

The EPA,[5] NBS,[6] and recent articles written by recognized leaders in the field of forensic analysis[4,7] are recommended sources of information for implementing a proper GC/MS protocol. We have adopted the suggested validation criteria from these references with our own experience of the scientific/legal arena of equine drug testing and forensic veterinary toxicology.[9] In addition, we have incorporated the guidelines given by the *Federal Register for Non-Clinical Laboratory Studies for Good Laboratory Practices* (GLP).[10]

Figure 1 shows a representative computer output from a so-called Mass Selective

Detector capillary GC/MS (Hewlett-Packard Co.) equipped with a dedicated data system (HP 5970A GC/MS Workstation).

The profiles of the ions m/z 69, 219, and 502 for the calibration compound, perfluorotributylamine (PFTBA), are shown in Figure 1 along with the plotted bar graph mass spectrum of PFTBA. Finally, the printed masses to two decimal places, the relative abundances, and isotopic ratios for each of the above-mentioned three ions are shown at the bottom of the same figure. These data document, on one page, all the relevant instrumental parameters utilized by the GC/MS system during the course of analyses performed on a particular day. This is an excellent way to record the "vital statistics" of the instrument since it represents a permanent record for the file. In addition, the "pulse" of the GC/MS system may be monitored on a continuing basis when these data are recorded daily and kept on file. Thus the GC/MS operator may monitor any deterioration of the performance of the instrument with time by reference to recorded files such as those shown in Figure 1. Equally important, however, is the value of having such instrument performance documentation associated with a positive drug finding. If the analyst cannot prove that the instrumentation utilized for proving the presence of an illegal substance was operated under acceptable conditions the analytical results could be deemed inadmissible. Thus, no matter how convincing the analytical results, the data may be considered invalid unless there is sufficient proof that the GC/MS was operating properly during the course of the analysis.

Historically, the choice of a suitable GC/MS test compound has been controversial. For a variety of reasons, researchers in differing areas have preferred specific test compounds for GC/MS depending on their particular needs or problems. Budde et al.[11] suggested the compound DFTPP as a suitable GC/MS test compound which has since been accepted by the EPA for uniform tuning criteria of GC/MS instruments in all EPA contract laboratories involved with environmental analyses. Although we admit there appears to be no ideal test compound for tuning purposes, we have adopted DFTPP into our GC/MS performance verification program because of its general suitability for such purposes.

We routinely introduce 15 ng of DFTPP through the GC/MS instrument subsequent to the tuning of the instrument and verify that we obtain the published mass spectrum for DFTPP on our instrumentation prior to any forensic GC/MS analysis. Particular care is taken to verify that the criteria established by Budde[11] are met with respect to relative abundance ratios and mass assignment. In this way we can be certain that the mass spectra will not vary from day to day and will agree with the mass spectra of other laboratories utilizing similar tuning criteria.

In summary, it is very helpful to document the entire experimental conditions utilized during the course of a forensic mass spectrometric analysis. The sample preparation history and all associated GC/MS conditions for the analysis of a naproxen administration equine urine are shown in Figure 2. This one-page summary conveniently summarizes all the relevant information needed. This information becomes part of the permanent record of the sample and is forensically very useful should the results be subjected to legal challenge.

The data shown in Figure 3 were obtained from the capillary GC/MS determination of 50 ng of DFTPP. Note that the characteristic ions at m/z 51, 69, 70, 127, 198, 275, 365, 441, 442, and 443 for this compound all are within the accepted relative abundance ranges.[11] This assures one that the ion relative abundances of compounds determined in subsequent experiments should correspond to those obtained in either previous or later analyses performed according to the same analytical procedure. This is particularly important when comparisons are made between data obtained from different quadrupole mass spectrometers which may be tuned differently according to the

```
EEEEEEEEEEEEEEEEEEEEEEEEEEEEEEEEEEEEEEEEEEEEEEEEEEEEEEEEEEEEEEEEEEEEEE
DDDDDDDDDDDDDDDDDDDDD EQUINE DRUG TESTING PROGRAM DDDDDDDDDDDDDDDDDDDDD
TTTTTTTTTTTTTTTTTTTTT     CORNELL UNIVERSITY     TTTTTTTTTTTTTTTTTTTTT
```

ANALYSIS: OSU-80 NAPROXEN ADMINISTRATION

DATE   : 03 Apr 85
PROCEDURE: TLC + EI GC/MS
ANALYST : D Silvestre

SAMPLE PREPARATION
        EXTRACTION METHOD: B.H.   VOLUME EXTRACTED: 9ml
        TLC SYSTEM   : 9:1   Rf   : .6
        ELUTION SOLVENT : ETOAC   FINAL SOLVENT   : ETOAC
                                           FINAL VOLUME   : 50ul

GC/MS INSTRUMENT: HP 5790A/5970A MSD

COLUMN: HP SPECIAL PURPOSE CAPILLARY
        LIQUID PHASE : Cross-linked Me Si. FILM THICKNESS: .50um
        COLUMN LENGTH: 17m   INT. DIAMETER : .20mm

GC CONDITIONS
        CARRIER GAS   : Helium   HEAD PRESSURE : 19psi
        INJECTION MODE: Grob splitless   PURGE OFF TIME: 1min
        INJECTOR TEMP : 260   INTERFACE TEMP: 285
        TEMP 1   : 130   TIME 1   : 1min
        TEMP RATE 1   : 15/min
        TEMP 2   : 270   TIME 2   : 10min
        TEMP RATE 2   :
        TEMP 3   :   TIME 3   :

        DERIVATIZATION: BSTFA (50% in ETOAC, .8ul injected)
        VOLUME OF SAMPLE INJECTED: 1ul

MS SCAN ACQUISITION PARAMETERS
        START TIME   : 5min
        LOW MASS   : m/z 40   HIGH MASS   : m/z 400
        SCAN THRESHOLD: 20   SCAN/SEC   : 1.2
        ELECTRON MULTIPLIER: 2000V

MS CALIBRATION DATA
        TUNING REPORT   (NEXT PAGE)
        DFTPP SPECTRUM (Data file: RUN356.D)

ANALYSIS DATA FILES
        RUN360.D: OSU-80 0HR 1ul/50ul +BSTFA
        RUN361.D: OSU-80 72HR 1ul/50ul +BSTFA
        RUN362.D: ETOAC 1ul +BSTFA
        RUN363.D: NAPROXEN 50NG +BSTFA
```

FIGURE 2.   Cover page for the analysis of a naproxen administration sample.
```

```
Data file : RUN356.D
Name info : DFTPP 50NG

Date       : 03 Apr 85
Analyst    : D. Silvestre
Instrument: HP 5790A/5970A MSD

Column     : HP Special Purpose Capillary
             .2mm/17m cross-linked Me Si
GC conditions
Carrier gas    : He        Head pressure : 19psi
Injection mode: Splitless  Purge off time: 1min
Injector Temp : 240        Interface Temp: 285
Temp 1         : 140       Time 1        : 1min
Temp rate 1    : 20/min
Temp 2         : 240       Time 2        : 5min
Temp rate 2    :
Temp 3         :           Time 3        :
MS scan acquisition parameters
Start time     : 3min
Low mass       : m/z 40    High mass     : m/z 450
Scan threshold: 20         Scan/sec      : 1
El. multiplier: 2000V
```

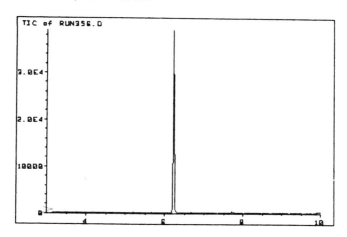

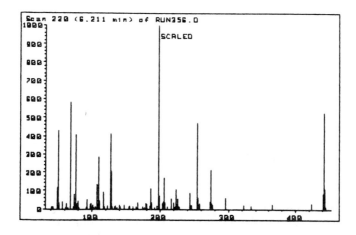

FIGURE 3.   GC/MS data for the determination of DFTPP.

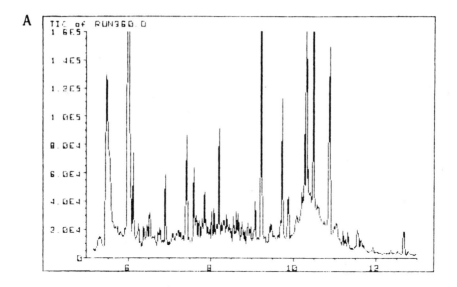

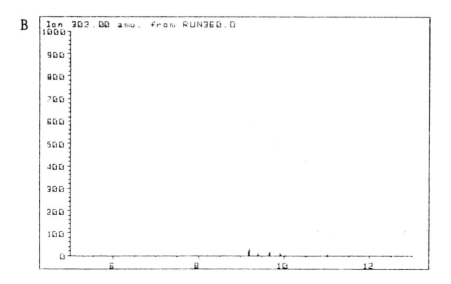

FIGURE 4.   (A) Total ion current (TIC) chromatogram from the GC/MS analysis of a
0-hr equine urine extract. (B) Extracted ion current profile (EICP) for the m/z 302 molecular
ion from the GC/MS analysis. (C) Mass spectrum of the 9.4-min retention time region.

procedures utilized by different operators. Thus the GC/MS data shown in Figure 3
document that the combined capillary GC/MS system is operating properly and that
the actual analysis of samples may commence.

Now we know that the entire GC/MS system is operating properly and that we may
begin the analysis in question. In order to preclude any possibility for contamination
of the "positive", or sample in question, one should always introduce the suspect
sample into the GC/MS first. In other words, the authentic standard of the suspected
foreign substance should not be introduced into the system until after the analysis of
the unknown sample. For added insurance that the data resulting from the unknown
sample will have no interferences, a sample extract obtained from a control, or "zero
hour", urine is analyzed first in a manner exactly the same as the unknown sample.

The upper portion of Figure 4 shows the total ion current chromatogram (TIC)

C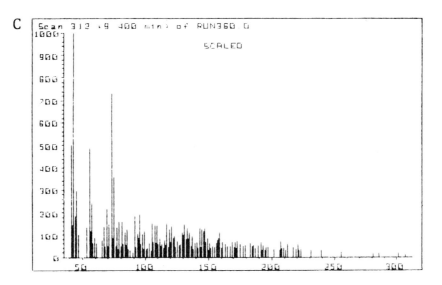

FIGURE 4C.

obtained from the capillary GC/MS analysis of a zero hour equine urine extract obtained by the same analytical preparation procedure as that used for the suspect sample. Figure 4B is the extracted ion current profile (EICP) for the m/z 302 ion, or molecular ion of the naproxen trimethylsilyl derivative (naproxen-TMS), while Figure 4C shows the mass spectrum recorded at the retention time expected for naproxen under these conditions. The data in Figure 4B and C document that there is no detectable naproxen interference in the sample preparation procedure and the GC/MS system immediately prior to the next injection which should be the suspect sample of interest.

Next, the suspected positive is analyzed in an identical manner as that described above. The corresponding GC/MS results for the analysis of the "positive" sample are given in Figure 5A to C. As shown in Figure 4A to C, this figure shows the TIC, EICP, and plotted mass spectrum for naproxen-TMS in the extract of an equine urine collected 72 hr subsequent to the administration of naproxen to a horse.

Although one could in principle simply compare the mass spectrum shown in Figure 5C to that obtained from authentic naproxen-TMS on a previous occasion or reference files, it is considered best to obtain the mass spectrum for authentic naproxen-TMS on the same instrument, on the same day, and under the same conditions as the sample in question.

III. EXAMPLES FROM SPORTS TESTING

A. Drug Testing in the 1984 Olympic Games

During the Olympic Games in Los Angeles, the presence of 162 banned substances and the quantitative determination of several more had to be carried out by the Olympic Drug Screening Laboratory. The Laboratory operated eight gas chromatographs, seven GD/MS systems, and one liquid chromatograph for this work.[12]

Hewlett-Packard model 5880 gas chromatographs were used for the screening of volatiles (e.g., barbiturates, alkaloids, opiates, and CNS stimulants such as amphetamine). If a positive sample was indicated by the appearance of a peak within a certain retention time window, a confirmation run would then be made on a model HP 5987 A GC/MS system.

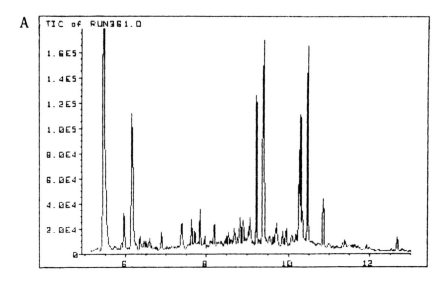

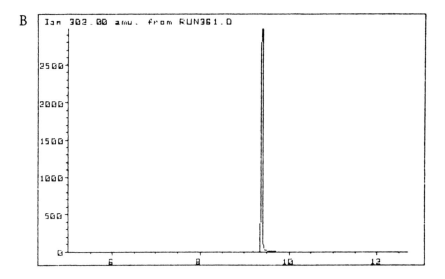

FIGURE 5. (A) TIC chromatogram from the GC/MS analysis of an extract from an equine urine collected 72 hr after the administration of naproxen. (B) EICP for the m/z ion from the GC/MS analysis. (C) Mass spectrum of the 9.4-min retention time region.

The HP 1090 liquid chromatograph was used to quantitate caffeine and to detect the presence of the stimulant, pemoline.

Screening for steroids, the major class of drugs abused by athletes, was the most challenging problem facing the laboratory. This type of analysis is not as straightforward as with other classes of drugs, since the body itself produces many similar types of compounds.

A simple chromatogram would be much too busy to interpret, even if sufficient levels were available for detection. To attain the specificity, as well as sensitivity, GC/MS in selected ion monitoring (SIM) mode was required. Instrument control and data reduction was carried out by dedicated computers.

The software procedure analyzed the data and printed a 2- to 18-page report (de-

C

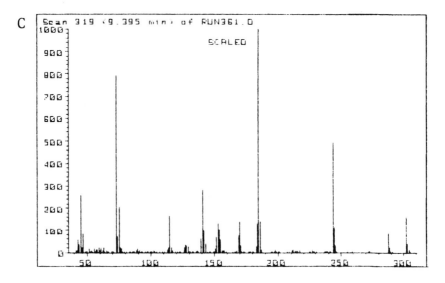

FIGURE 5C.

pending on the number of positives) with numeric results as well as graphics representation for visual confirmation. An example of the type of data acquired is shown in Figure 6.

B. Racehorse Drug Testing

1. GC/MS

Levallorphan is a narcotic antagonist drug that may be administered to racehorses in certain instances. Its determination in equine urine is hindered by its relatively low level in urine volumes collected several hours subsequent to administration, and the chemical complexity of equine urine. Liquid-liquid extraction of the parent drug from urine followed by preparative TLC readily prepares a urine extract for subsequent capillary GC/MS analysis of the sample. Preparative TLC facilitates removal of the high levels of endogenous materials which may overload the capillary GC column.

Figure 7A to C shows the capillary GC/MS analysis of an equine urine extract containing levallorphan. Figure 7A shows the total ion current (TIC) for the 5- to 9-min time window of this capillary GC/MS experiment while Figure 7B shows the ion current profile for m/z 335 for the TMS derivative of levallorphan and Figure 7C shows the mass spectrum for this compound which eluted at 6.174 min. The data obtained from this sample was compared with the GC/MS data from a standard which corroborated the identity of levallorphan in the urine sample.

2. LC/MS

Stereoselectivity is an important phenomenon in biological systems. In most cases, optical enantiomers exhibit different biological activities. The stereoisomers of ibuprofen, a popular nonsteroidal anti-inflammatory agent, show significant differences in potency and metabolism. LC/MS with a chiral stationary phase (CSP/LC/MS) was used for the detection and identification of ibuprofen enantiomers in equine urine after administration of (S,R)-ibuprofen.[13] The preparative TLC scrape was derivatized to form the benzylamide and analyzed by CSP/LC/MS. The resulting ion current profile is shown in Figure 8B and compared with the corresponding data obtained from an authentic mixture of (R,S)-ibuprofen benzylamide shown in Figure 8A. The smaller chromatographic peak observed in Figure 8B at 6.1 min is the (R)-(−)-ibuprofen en-

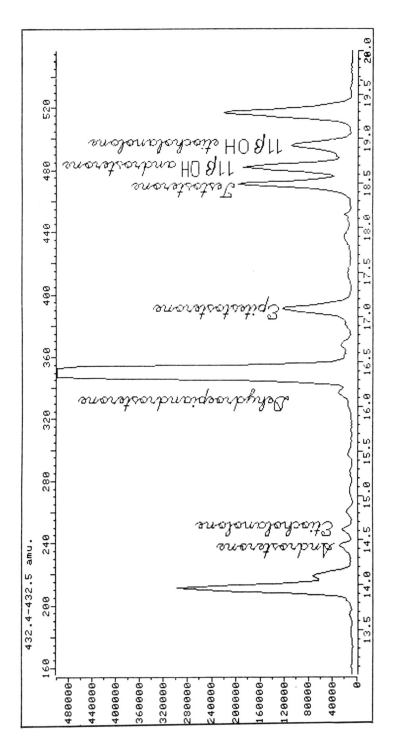

FIGURE 6. Steroid screen ion chromatogram printout. (Reprinted with permission from R. K. Latven, Hewlett-Packard).

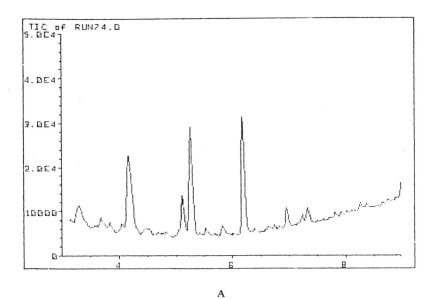

A

B

FIGURE 7. Capillary GC/MS analysis of an equine urine extract containing levallorphan. (A) TIC for the GC/MS analysis. (B) Ion current profile for m/z 355 ion. (C) Mass spectrum of eluted compound at 6.174 min.

antiomer by comparison with the corresponding peak in Figure 8A. The CSP/LC/MS mass spectra shown in Figure 9 were obtained from the ion current profiles for authentic (R,S)-ibuprofen benzylamide shown in Figure 8A. These data clearly show the abundant (M + 1) ion of ibuprofen benzylamide. In addition a m/z 206 fragment, possibly resulting from the fragmentation of ibuprofen benzylamide to yield parent ibuprofen, is present. The similarity of the two mass spectra shown in Figure 9 reveals the difficulty of distinguishing these enantiomers by their mass spectra alone. However, the correlation of the mass spectra with the respective CSP/HPLC retention times makes identification of the optical isomers possible.

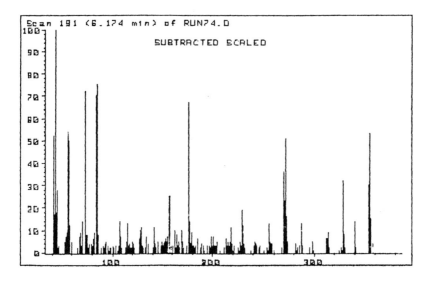

FIGURE 7C.

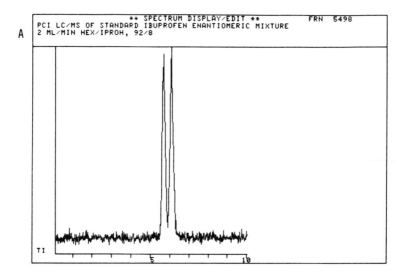

FIGURE 8.   (A) LC/MS ion current profile of standard mixture of *(R,S)*-ibupro-
fen benzylamide. (B) LC/MS ion current profile of equine urine ibuprofen TLC
scrape. (From Crowther, J. B., et al., *Anal. Chem.*, 56, 2921, 1984. With permission.)

Another example is the use of micro-LC/MS for the analysis of equine urine.[3] A
comparison between a micro-HPLC/UV chromatogram of an equine urine extract
containing betamethasone and the corresponding micro-LC/MS ion current chromatograms (Figure 10) shows the similarity between these two data sets. The major advantage of the ion current chromatograms is the ability to "extract" the ion current
profile of the major metabolite, 6-β-hydroxybeta-methasone from the envelope of coe-
luting endogenous compounds. A minor metabolite was detected at 10.2-min retention
time from the TIC, which is not evident in the UV chromatogram.

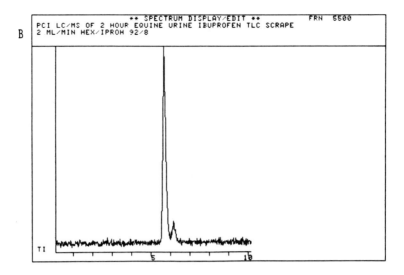

FIGURE 8B.

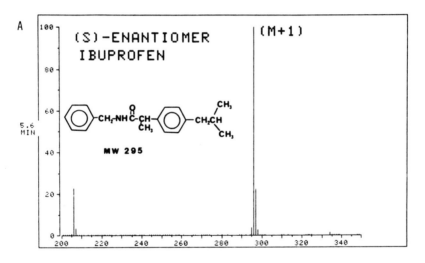

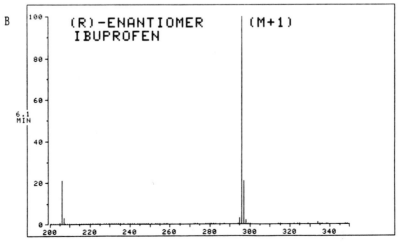

FIGURE 9. LC/MS mass spectra of ibuprofen enantiomers. (From Crowther, J. B., et al., *Anal. Chem.*, 56, 2921, 1984. With permission.)

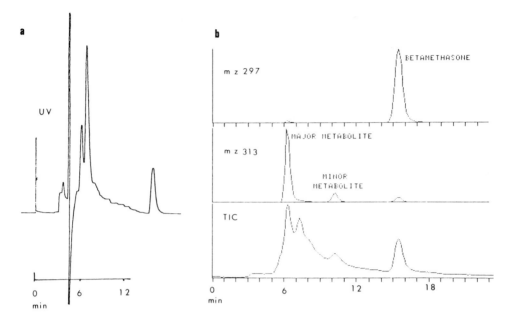

FIGURE 10. Micro-HPLC (a) and micro-LC/MS ion current (b) chromatograms of an equine urine extract containing betamethasone. (From Lee, E. D. and Henion, J. D., *J. Chromatogr. Sci.,* 23, 253, 1985. With permission.)

3. Tandem Mass Spectrometry (MS/MS)

Clenbuterol is a potent respiratory stimulant administered to horses in doses ranging from 100 to 200 μg. The rapid onset of pharmacological effects coupled with a short duration are manifested by low levels of the parent drug in the plasma and urine. The detection of clenbuterol in equine urine and its confirmation by GC/MS can be accomplished by scrupulous preparative TLC procedures and derivatization prior to GC/MS analysis on well-conditioned GC columns. A simplified confirmation step requiring less sample clean-up is therefore desirable due to sample losses resulting from multistep cleanup.

MS/MS has been shown to be a highly specific and sensitive technique for the analysis of trace components in complex samples. MS/MS has been used for the detection and identification of drugs in equine urine, using a TAGA® 6000 E triple quadrupole mass spectrometer equipped with an atmospheric pressure ionization (API) source.[14] Figure 11A shows the full-scan API mass spectrum obtained from an eluted TLC scrape of an equine urine extract obtained after intravenous administration of 200 μg of clenbuterol. Although there is a relatively small $(M + 1)^+$ ion at m/z 277, there are many other abundant ions present that preclude identification of clenbuterol from this mixed mass spectrum. The collisionally activated dissociation (CAD) mass spectrum of m/z 277, as shown in Figure 11B, agrees well with the CAD spectrum obtained from standard clenbuterol, and makes therefore positive identification of clenbuterol possible.

ACKNOWLEDGMENT

We thank the New York State Racing and Wagering Board and the racing industry of New York State for financial support of our work.

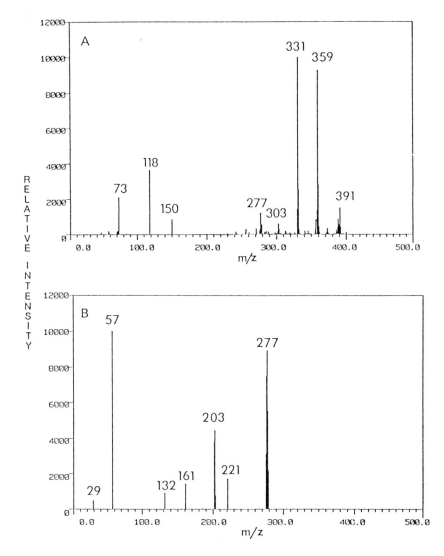

FIGURE 11. (A) Full-scan API mass spectrum of an eluted TLC scrape of an equine urine extract subsequent to the administration of clenbuterol. (B) Collisionally activated dissociation (CAD) spectrum of the m/z 277 ion. (From Henion, J. D., et al., *J. Chromatogr.*, 271, 107, 1983. With permission.)

REFERENCES

1. Rostaing, B. and Sullivan, R., Triumphs tainted with blood, *Sports Illustrated*, January 21, 1985, 12.
2. McFadden, W., *Techniques of Combined Gas Chromatography/Mass Spectrometry: Application in Organic Analysis*, John Wiley & Sons, New York, 1973.
3. Lee, E. D. and Henion, J. D., Micro-liquid chromatography/mass spectrometry with direct liquid introduction, *J. Chromatogr. Sci.*, 23, 253, 1985.
4a. Brettell, T. A. and Saferstein, R., Forensic science, *Anal. Chem.*, 55, 192, 1983.
4b. Sobol, S. P., Forensic drug analysis, *Am. J. Forensic Med. Pathol.*, 3, 93, 1982.
5. Budde, W. L. and Eichelberger, J. W., Performance tests for the evaluation of computerized GC/MS equipment and laboratories, EPA 6001 4-80 025, April, 1980.
6. American Society for Testing Materials, Section 14, Vol. 14.01, E. 899.82, 808.

7. Taylor, J. K., Validation of analytical methods, *Anal. Chem.,* 55, 600A, 1983.

8. Keith, L. H. et al., Principles of environmental analysis, *Anal. Chem.,* 55, 2210, 1983.

9. Covey, T. R. and Henion, J. D., High performance liquid chromatography in veterinary toxicology, *J. Liq. Chromatogr.,* 7(S-2), 205, 1984.

10. F.D.A., Federal Register, Nonclinical Lab Studies Good Laboratory Practice Regulations, Part II, Food and Drug Administration, Department of Health, Education, and Welfare, December 22, 1978.

11. Eichelberger, J. W., Harris, L. W., and Budde, W. L., *Anal. Chem.,* 47, 995, 1975.

12. Latven, R. K., personal communication.

13. Crowther, J. B., Covey, T. R., Dewey, E. A., and Henion, J. D., Liquid chromatographic/mass spectrometric determination of optically active drugs, *Anal. Chem.,* 56, 2921, 1984.

14. Henion, J. D., Maylin, G. A., and Thomson, B. A., Determination of drugs in biological samples by thin-layer chromatography-tandem mass spectrometry, *J. Chromatogr.,* 271, 107, 1983.

Chapter 4

MASS SPECTROMETRY OF EXPLOSIVES

Jehuda Yinon

TABLE OF CONTENTS

I. INTRODUCTION

Mass spectrometry (MS) has become a widely used tool for both identification and detection of explosives.

Forensic identification of post-explosion residues in debris material is of major interest in the criminalistic investigation of a bombing because it will help in finding the link between the explosive used and the suspect. The result of the analysis is also needed as judicial evidence in court. The unambiguous determination of the explosive is often complicated by the small quantities of material available after separation from other substances present in the debris of the bombing scene. The sensitivity and specificity of the mass spectrometer make it a suitable technique for this application.

The detection of concealed explosives in airline baggage and in mail needs a sensitive, fast, and reliable analytical device. The mass spectrometer, having these features, has been used as a "sniffing detector" in various configurations.

Although MS of explosives has been reviewed previously,[1] it is our aim in this chapter to emphasize the analytical aspects of the use of MS for explosives.

In the earlier stages of forensic MS, electron impact (EI) was used for the analysis of explosives. In EI, where the electron energy is about 70 eV, extensive fragmentation of the analyzed molecules occurs, and sometimes no molecular ion is obtained. Fragmentation patterns can be correlated with specific functional groups, thus enabling the recognition of many structural features in molecules. A fragmentation pattern of a molecule can therefore be considered a "fingerprint" which can be used as an identification tool. Libraries of mass spectra of explosives have been compiled by various laboratories and can be used as references to compare with unknown mass spectra. EIMS is not suited for the analysis of mixtures because the resulting mass spectra are usually too complex for identification. The combination of gas chromatography/mass spectrometry (GC/MS) can be used for the analysis of explosive mixtures only to a limited extent due to the thermal instability of the explosives which may cause the decomposition of the compounds in the heated GC. However, the use of chemically inert fused silica capillary columns improves the performance of GC/MS analysis of explosives. The discovery of chemical ionization (CI) as a soft ionization technique brought an increased interest in the application of MS to forensic identification of explosives. The CI mass spectrum, providing molecular weight information of the investigated sample, is being complemented by the EI fragmentation pattern to produce a reliable identification technique.

Many additional ionization techniques have been used for both identification and detection, including negative-ion mass spectrometry, atmospheric pressure ionization (API), field ionization (FI), and field desorption (FD). Two new analytical techniques, high-performance liquid chromatography (HPLC) in combination with CIMS (LC/MS) have also been used for the analysis and characterization of explosives.

II. IDENTIFICATION OF EXPLOSIVES

A. Aromatic Nitro Compounds

One of the most widely used nitroaromatic explosives is 2,4,6-trinitrotoluene (TNT). The EI mass spectrum of TNT (Figure 1) was recorded by several groups using low-resolution[2-5] and high-resolution[6-8] MS. The most abundant ion in the EI mass spectrum of TNT is the $(M - OH)^+$ ion at m/z 210 formed by the loss of a hydroxyl radical from the molecular ion due to an *ortho* effect.[9] Other characteristic ions are $(M - 2OH)^+$ at m/z 193, $(M - HNO_2)^+$ at m/z 180, $(M - 3NO_2)^+$ at m/z 89, and NO^+ at m/z 30.

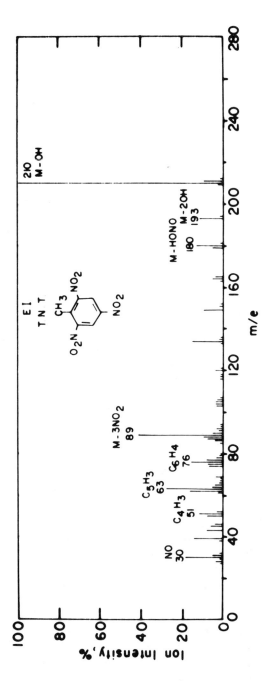

FIGURE 1. EI mass spectrum of TNT. (From Frigerio, A. and Castagnoli, N., Eds., *Advances in Mass Spectrometry in Biochemistry and Medicine*, Vol. 1, Spectrum Publications, New York, 1976, 369. With permission.)

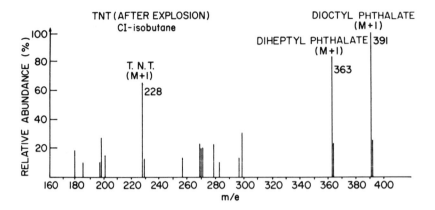

FIGURE 2. CI mass spectrum of a post-explosion residue extract. (From Yinon, J. and Zitrin, S., *J. Forensic Sci.*, 22, 742, 1977. With permission.)

The detection limit of TNT using a GC/MS system with a fused silica capillary column was found to be 50 pg.[10] This was obtained by EI ionization and single ion monitoring of the ion at m/z 210. High-sensitivity detection of trace amounts of explosives is very important in post-explosion residues.

The CI mass spectrum of TNT contains mainly the MH$^+$ ion at m/z 228. The CI spectrum of TNT has been recorded with various reagents: hydrogen,[11] isobutane,[4,12] methane,[4,5,13,14] ammonia,[13] and water.[15,16]

Figure 2 shows the CI mass spectrum of a sample taken from the debris of an explosion site.[17] The sample was extracted with acetone and, after evaporation of the solvent, the residue was introduced with the solid probe into the mass spectrometer. Isobutane was used as CI reagent gas. The peak at m/z 228 can be clearly identified as the MH$^+$ ion of TNT. Abundant phthalate-ester ions were observed at m/z 363 and m/z 391 due to diheptyl phthalate and dioctyl phthalate, originating from the plastic garbage container in which the bomb had been placed. The mass spectrum also contained additional peaks which were produced from impurities in the debris.

When using water as reagent, an additional ion is produced, (MH − 30)$^+$ at m/z 198, which is due to the reduction of the MH$^+$ ion to its corresponding amine.[15] The relative abundance of this ion can be as high as or even higher than that of the MH$^+$ ion, depending on the pressure of water vapor in the ion source and on the source temperature.

As many explosives are heat-labile compounds, they should be analyzed at the lowest possible temperature while producing analytically useful mass spectra. In direct-exposure CI, the sample is in direct contact with the ionizing plasma. Therefore, protonated molecular ions are produced at lower temperatures. Such a direct-exposure CI mass spectrum of TNT with isobutane as reagent gas was obtained[18] at a source temperature of 75°C.

Low-pressure negative-ion mass spectra are highly dependent on electron energy because of the variety of processes involved in their formation. Therefore, the analytical use of this technique is limited. The dominant ion in the low-pressure, negative-ion mass spectrum of TNT at 6 eV electron energy[19] is NO$_2^-$. This ion carries 57% of the total ion current. At 2 eV electron energy, the mass spectrum consists of the NO$_2^-$ ion in addition to a number of structurally significant ions such as (M − OH)$^-$ at m/z 210, (M − NO)$^-$ at m/z 197, (M − NO$_2$)$^-$ at m/z 181, (M − 2NO)$^-$ at m/z 167, and (M − NO$_2$ − NO)$^-$ at m/z 151.

High-pressure negative-ion MS, due to its higher sensitivity, has been found to be

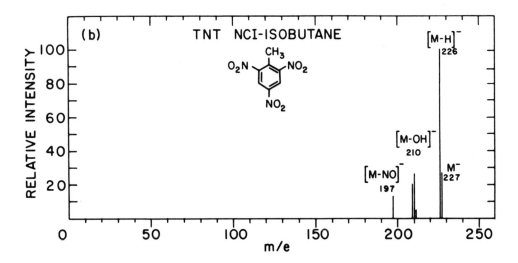

FIGURE 3. NCI mass spectrum of TNT. (From Yinon, J., *J. Forensic Sci.,* 25, 401, 1980. With permission.)

useful in the analysis of explosives. The negative ion chemical ionization (NCI) mass spectrum of TNT[20] with isobutane as moderator gas at a pressure of 0.2 to 0.4 torr is shown in Figure 3. The base peak at m/z 226 is due to the (M − H)⁻ ion formed by proton transfer. Other major ions are the molecular ion M⁻ at m/z 227, (M − OH)⁻ at m/z 210, and (M − NO)⁻ at m/z 197.

TNT has been analyzed by GC/MS with a fused silica capillary column using NCI with methane as moderator gas.[10] The detection limit was found to be 125 ng, obtained by single ion monitoring of the ion at m/z 210.

Negative-ion atmospheric pressure ionization (API) has been used[21] for trace analysis identification of TNT. Resonance electron capture was the major process of ionization. The detection limit for TNT, using a direct probe, and monitoring the M⁻ ion at m/z 227, was found to be 50 fg,[21] based upon a signal-to-noise ratio of 3:1.

The field desorption (FD) mass spectrum of TNT[22] consists of the molecular ion M⁺ as base peak and of the (M + H)⁺ and (M + 2H)⁺ ions.

Crude TNT also contains trinitrotoluenes other than the 2,4,6 isomer, as well as some isomeric dinitrotoluenes. The purity of the product is determined by the melting point, the minimum value for high-grade TNT being 80.2°C. Unless adequate purity is achieved, slow exudation of impurities can occur during storage and the TNT then becomes insensitive. It is therefore important to know the mass spectra of the isomeric trinitrotoluenes and dinitrotoluenes. The EI mass spectra of the isomeric trinitrotoluenes[23] with the exception of the 3,4,5 isomer contain an abundant (M − OH)⁺ ion at m/z 210 which in most cases forms the base peak. The 3,4,5 isomer is the only one without an *ortho* methyl group, and has therefore no ion at m/z 210 in its spectrum. The (M − OH)⁺ ion subsequently loses NO and NO₂, producing ions at m/z 180 and m/z 134, with the exception of the 2,4,5 isomer. The mass spectra of all isomers contain a significant molecular ion M⁺, with the exception of the 2,4,6 in which the M⁺ is very small.

The CI mass spectra of the 2,4,5 and 2,3,6 isomers,[5] as recorded with methane as reagent, are similar to the one of the 2,4,6 isomer.

The dinitrotoluene (DNT) isomers[5] having a nitro group in the 2-position lose a neutral OH fragment from the parent ion, due to an *ortho* effect, to form a highly abundant ion at m/z 165, which in most cases is the base peak. 2,6-DNT loses two successive OH fragments to form an ion at m/z 148. 2,3-DNT has an abundant ion at m/z 135 due to $(M - OH - NO)^+$ forming the base peak. The 3,5 and 3,4 isomers have an abundant molecular ion M^+ at m/z 182, which in 3,5-DNT forms the base peak. In 3,4-DNT the NO^+ ion forms the base peak. The CI-methane mass spectra of all six DNT isomers contain mainly the MH^+ ion at m/z 183 which forms the base peak.

The EI mass spectra of picric acid (2,4,6-trinitrophenol),[6,24] picryl chloride (2,4,6-trinitrochlorobenzene),[2,6] and picramide (2,4,6-trinitroaniline)[6] are characterized by abundant molecular ions M^+ and $(M - 3NO_2)^+$ and NO^+ fragment ions. In the mass spectrum of picryl chloride the $(M - 3NO_2 - Cl)^+$ ion is also highly abundant. The EI mass spectra of ammonium picrate[6] and 2,4,6-trinitrophenetole (TNP) are very similar to those of picric acid. Ammonium picrate decomposes thermally to picric acid. In the EI spectrum of TNP the most abundant ion is the $(M - C_2H_4)^+$ ion, which is identical to the molecular ion of picric acid. This ion decomposes further by a fragmentation pattern similar to that of picric acid, which explains the similarity between the two mass spectra. The EI mass spectra of 2,4,6-trinitro-*m*-xylene (TNX)[6] and 2,4,6-trinitro-*m*-cresol (TNC)[6] are characterized by abundant $(M - OH)^+$ and by $(M - 2OH)^+$ ions resulting from an *ortho* and a double *ortho* effect, respectively. In the EI mass spectrum of TNX an $(M - 3OH)^+$ ion is observed, due to a triple *ortho* effect.

The CI mass spectra with methane and isobutane as reagents of picric acid, picryl chloride, picramide, TNX, and TNC contain mainly the MH^+ ion which forms the base peak. In the CI-isobutane mass spectrum of TNP the base peak is formed by the MH^+ ion, while in its CI-methane spectrum it is formed by the $(MH - C_2H_4)^+$ ion at m/z 230.

The API mass spectrum of TNP[26] injected into the ion source as a solution with cyclohexanone, using nitrogen at atmospheric pressure as carrier gas, consists of one single ion, MH^+, at m/z 258. When water is used as reagent, the CI mass spectra[15] of picramide, picric acid, and TNC contain highly abundant $(MH - 30)^+$ ions which are formed by a reduction process occurring in the CI source.

Hexanitrostilbene (HNS, 1) is an explosive which has found increasing application because of its low sensitivity to shock, friction, and impact and good stability at elevated temperatures.

1

The EI mass spectrum of HNS[27] has a base peak at m/z 30 due to the NO^+ ion and highly abundant ions at m/z 44 $(N_2O)^+$, m/z 119 $(C_6HNO_2)^+$, and m/z 167 $(C_6H_3N_2O_4)^+$. The highest mass ion appears at m/z 240 due to $(C_8H_6N_3O_6)^+$.

The low-pressure, negative-ion mass spectrum of HNS[27] has its base peak at m/z 212 $(C_6H_2N_3O_6)^-$ and a highly abundant molecular ion M^- at m/z 450. Fragment ions appear at m/z 404 $(M - NO_2)^-$, m/z 387 $(M - HNO_3)^-$, m/z 375 $(M - CHNO_3)^-$, m/z 358 $(M - 2NO_2)^-$, and m/z 46 $(NO_2)^-$.

The CI mass spectrum of HNS with ammonia as reagent[27] is similar to its EI spectrum.

The NCI mass spectrum of HNS with carbon tetrachloride as reagent[27] consists mainly of the molecular ion M^- at m/z 450.

B. Nitrate Esters

Glycerol trinitrate or nitroglycerin (NG) is one of the most widely used explosives as a component in dynamite and in mining explosives. Pentaerythritol tetranitrate (PETN) is a very powerful explosive. Because of its high sensitivity to friction and impact, it is usually mixed with other explosives or with plasticized nitrocellulose to form a plastic explosive.

The EI mass spectra of nitrate esters are characterized by an abundant NO_2^+ ion and no molecular ion.[28] It is therefore difficult to identify these explosives by EIMS.

The CI mass spectrum of ethylene glycol dinitrate (EGDN)[14] with methane as reagent has its base peak at m/z 46 $(NO_2)^+$, a highly abundant ion $(MH - HNO_3)^+$ at m/z 90, an MH^+ ion at m/z 153 and adduct ions $(M + NO)^+$ at m/z 182, and $(M + NO_2)^+$ at m/z 198. These adduct ions are a result of ion-molecule reactions in the CI source between the fragment ions NO^+ and NO_2^+ and the sample molecules.

The CI mass spectrum of diethylene glycol dinitrate (DEGN) with water as reagent[29] is shown in Figure 4. The base peak is at m/z 197 (MH^+) and a major fragment ion, $(MH - CH_2OCH_2)^+$, appears at m/z 153.

Due to its high volatility and thermal instability, NG produces CI spectra[1,13,14,18,29] which depend strongly on the reagent gas, its pressure in the ion source, and the sample and source temperatures. MH^+ ions are produced at various relative abundances, from 2 to 100%. The CI spectrum with water as reagent[29] produces a base peak at m/z 228 $(MH)^+$ and an abundant fragment ion $(MH - CH_3NO)^+$ at m/z 183; with isobutane,[18] NG produces a base peak at m/z 183, an MH^+ ion having an abundance of 60%, and a series of adduct ions; with ammonia,[13] NG produces a base peak at m/z 76 $(CH_2ONO_2)^+$ and with methane[13,14] a base peak at m/z 46 $(NO_2)^+$. In all the CI mass spectra of NG an abundant $(MH - HNO_3)^+$ ion at m/z 165 is formed.

NCI mass spectra of NG with isobutane as moderator gas have been obtained[20] at source and probe temperatures of 80 and 25°C, respectively. Major fragment ions observed are NO_2^- and ONO_2^- (which forms the base peak at m/z 62). Additional fragment ions observed are $O(NO_2)_2^-$ at m/z 108, $(ONO_2)_2^-$ at m/z 124, $(M - NO_2)^-$ at m/z 181, and $(M - NO)^-$ at m/z 197, as well as a molecular ion M^- at m/z 227 and two adduct ions $(M + O)^-$ at m/z 243 and $(M + ONO_2)^-$ at m/z 289.

CI mass spectra of PETN have been recorded with various reagents, at different conditions.[5,11,12,18,30,31] The obtained mass spectral ions are recorded in Table 1.

In the NCI mass spectrum of PETN with isobutane as moderator gas,[20] the base peak is at m/z 62, $(ONO_2)^-$, and a major fragment ion is $(NO_2)^-$ at m/z 46. Additional fragment ions are $(M - NONO_2)^-$ at m/z 240, $(M + H - NONO_2)^-$ at m/z 241, and $(M - NO)^-$ at m/z 286. No molecular ion is observed, instead $(M + H)^-$ and $(M - H)^-$ ions are observed. Adduct ions $(M - H + O)^-$ at m/z 331, $(M + OH)^-$ at m/z 333, $(M + NO_2)^-$ at m/z 362, and $(M + ONO_2)^-$ at m/z 378 are formed.

The field ionization (FI) mass spectrum of PETN[32] includes the following ions: M^+ at m/z 316 with a relative abundance of 80%, $(M - NO_2)^+$ at m/z 270 (35%), NO_2^+ at m/z 46 (100%), NO^+ at m/z 30 (39%), and O^+ at m/z 16 (11%).

The FD mass spectrum of PETN[22] has a base peak at m/z 240 due to $(M - CH_2ONO_2)^+$ and the following highly abundant ions: MH^+ at m/z 317, $(MH - NO_2)^+$ at m/z 271, $(M - NO_3)^+$ at m/z 254, $(M - NO_2 - CH_2ONO_2)^+$ at m/z 194, $CH_2ONO_2^+$ at m/z 76, NO_3^+ at m/z 62, and NO_2^+ at m/z 46.

C. Nitramines

The most important nitramine explosives are 2,4,6,N-tetranitro-N-methylaniline (tetryl), 1,3,5-trinitro-1,3,5-triazacyclohexane (hexogen or RDX), and 1,3,5,7-tetranitro-1,3,5,7-tetraza-cyclooctane (octogen or HMX).

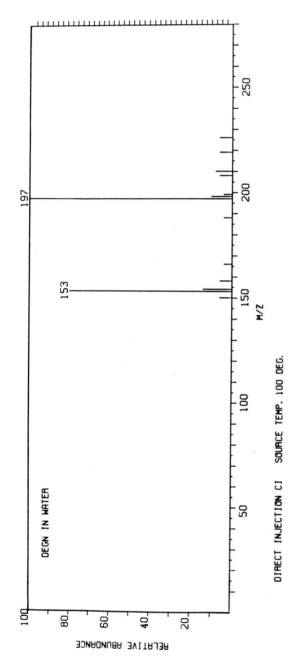

FIGURE 4.   CI-water mass spectrum of DEGN.

Table 1
CI MASS SPECTRAL IONS OF PETN

	Relative abundance (%)					
	GC inlet at 200°C; source 150°C methane 1 torr	Probe 250°C; source 110°C isobutane 0.2 torr	Direct exposure; source 75°C isobutane 0.15 torr	Probe 160°C; source 170°C water 0.1 torr	Probe 100°C; source 120°C hydrogen 1 torr	Probe 120°C; source 150°C ammonia 0.5—1.0 torr
Reference Ion	5	12	18	30	11	31
NO^+						18
NO_2^+				50	10	
					15	
						8
$(NO_2 + H_2O)^+$				45		
	65					
	35					
	22					
		40				
		8				
	100					
						5
	70					
	22					
						9
						35
	70					
	35	4				
$(2CH_2ONO_2 + NO)^+$				18		
$(244 - NO_2 + H)^+$						10
	30					
		4				
$(289 - NO_2 + H)^+$						23
$(MH - HNO_3)^+$	57	13	18			
				18		
	12	12				
	35	10			9	
$(M + NH_4 - NO_2 + H)^+$						100
		15	23			
MH^+	58	25	45	70	75	
$(M + NH_4)^+$						55
$(M + NO)^+$		14	10	9	5	
$(M + C_3H_7)^+$		35	100			
$(M + NO_2)^+$		100	42	100	10	

Figure 5 shows the EI mass spectra of tetryl[5] at electron energies of 70 and 20 eV. This example demonstrates the analytical application of reducing the electron energy, which in some cases will produce a mass spectrum containing high mass ions with higher relative abundances. Such a spectrum is much more meaningful for identification purposes. In the 20-eV mass spectrum the base peak is at m/z 241 formed by the $(M - NO_2)^+$ ion.

The CI mass spectrum of tetryl has been recorded by various groups.[4,5,14,16,18,25] Figure 6 shows the direct-exposure CI-isobutane mass spectra of tetryl at two different source pressures.[18] The relative abundances of the ions at m/z 242, m/z 243, and m/z 288, $(MH)^+$ depend on the source pressure.

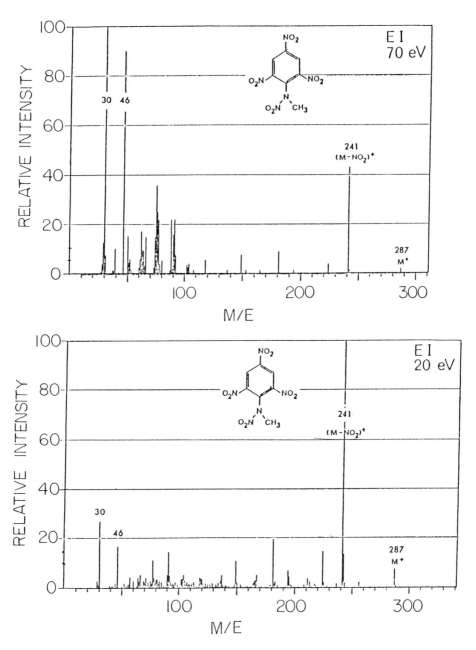

FIGURE 5. EI mass spectra of tetryl at 70 and 20 eV. (From Alm, A. et al., FOA Report C20267-D1, National Defense Research Institute, S-104 50 Stockholm, Sweden.)

The NCI mass spectrum of tetryl with isobutane as moderator[33] is mainly similar to its EI mass spectrum at 20 eV: the base peak, formed by the $(M - NO_2)^-$ ion, is at m/z 241.

The FD mass spectrum of tetryl[22] consists of a base peak ion at m/z 287 (M^+), adduct ions at m/z 288, 289, 290, and fragment ions at m/z 271 $(M - O)^+$, 241 $(M - NO_2)$, 225 $(M - ONO_2)^+$, 74 $(CH_2NNO_2)^+$, and a whole series of unidentified ions. Because the emitter heating current has a major influence on the mass spectrum, it is important to control the heating current in order to obtain reproducible mass spectra.

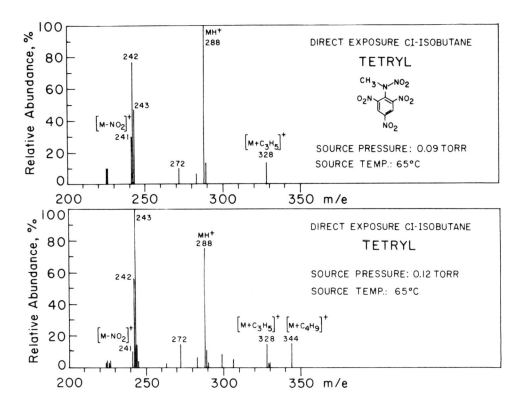

FIGURE 6. Direct exposure CI-isobutane mass spectra of tetryl at two different source pressures. (From Yinon, J., *Org. Mass Spectrom.*, 15, 637, 1980. With permission.)

RDX is one of the most important military explosives. The EI mass spectra of RDX as recorded by different groups[2,4,5,34-36] differ in the relative abundances of the major ions, which are mostly in the low-mass range. This is due to differences in thermal conditions (source and probe temperatures) which cause various degrees of thermal decomposition of this explosive.

The CI mass spectrum of RDX is sensitive to the type of reagent gas, to its pressure in the ion source, and to the source temperature. The CI mass spectral ions of RDX recorded by various groups under different conditions have been summarized.[1] An example is given in Figure 7 which shows the direct exposure CI isobutane mass spectra of RDX at two different source temperatures:[18] 75 and 90°C. At 75°C the MH$^+$ ion at m/z 223 forms the base peak and several adduct ions are observed. At 90°C these adduct ions disappear, but the relative abundance of the lower mass ions increases.

The API mass spectrum of RDX with nitrogen as carrier gas and cyclohexanone as solvent[26] consists of three major ions: $(M + H - CH_2NNO_2)^+$ at m/z 149 (100%), $(M + H - HNO_2)^+$ at m/z 176 (50%), and $(M + H)^+$ at m/z 223 (75%).

The NCI mass spectrum of RDX with isobutane as moderator gas[20] produces adduct ions such as $(M + H)^-$, $(M + NO)^-$, and $(M + NO_2)^-$ as well as fragment ions such as $(M - H)^-$, $(M - NO_2)^-$, $(M - CH_2NNO_2)^-$, $(M - NO_2HNO_2)^-$, $(NO_2HNO_2)^-$, and $(NO_2)^-$. The NCI mass spectrum of RDX with methane or ammonia as moderator gas[33] produced a base peak at m/z 490, probably due to the adduct ion $(2M + NO_2)^-$.

The FD mass spectrum of RDX[22] consists of the following major ions: $(MH)^+$ at m/z 223, $(MH - H_2O)^+$ at m/z 205, $(M - NO_2)^+$ at m/z 176, $(M - 2HNO_2)^+$ at m/z 128, $(C_3H_6N_3O_2)^+$ at m/z 116, $(C_2H_4N_3O_2)^+$ at m/z 102, $(CH_2NNO_2H)^+$ at m/z 75, $(CH_2NNO_2)^+$ at m/z 74, and two unidentified ions at m/z 82 and 83.

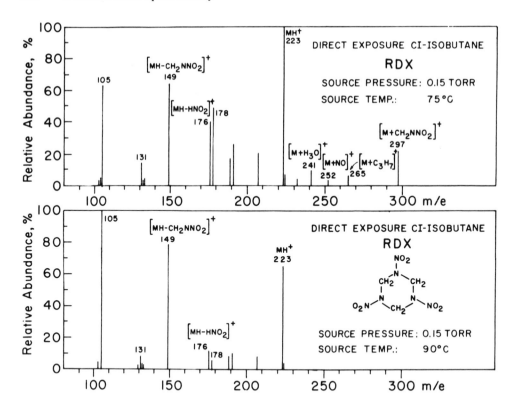

FIGURE 7. Direct exposure CI-isobutane mass spectra of RDX at two different source temperatures. (From Yinon, J., *Org. Mass Spectrom.*, 15, 637, 1980. With permission.)

HMX is used as an explosive and as a propellant for solid fuel rockets. The EI mass spectrum of HMX has been recorded by several groups.[2,4,5,36] Like RDX, HMX undergoes ring contraction by elimination of CH_2NNO_2, producing an ion at m/z 222. This ion has a much higher abundance in the EI mass spectrum of HMX than in that of RDX and therefore can be used to differentiate between RDX and HMX. Common fragment ions in both compounds are $(C_3H_5N_6O_5)^+$ at m/z 205, $(C_3H_6N_5O_4)^+$ at m/z 176, and $(C_2H_4N_4O_4)^+$ at m/z 148, as well as fragment ions formed from these ions. Both EI spectra of RDX and HMX are characterized by losses of CH_2NNO_2, CH_3NNO_2, NO_2, NO, HNO_2, and HNO.[36]

The major fragment ions in the CI mass spectrum of HMX are similar to those of RDX. An additional fragment ion is $(MH - CH_2NNO_2)^+$ at m/z 223 which is isomeric with MH^+ of RDX. Figure 8 shows a typical CI-methane spectrum of HMX,[13] recorded at a source and probe temperature of 100°C.

The NCI mass spectrum of HMX[20,33,36] produces adduct ions $(M + NO)^-$ and $(M + NO_2)^-$ and major fragment ions $(M - CH_2NNO_2)^-$ at m/z 222, $(C_3H_6N_5O_4)^-$ at m/z 176, $(M - 2CH_2NNO_2)^-$ at m/z 148, $(C_3H_5N_4O_2)^-$ at m/z 129, $(C_2H_4N_3O_2)^-$ at m/z 102, and $(NO_2)^-$ at m/z 46.

The FD mass spectrum of HMX[22] is shown in Figure 9. It includes an abundant MH^+ ion at m/z 297 and a series of characteristic fragment ions which can be used for identification of the compound. For example, the fragment ions $(2CH_2NNO_2 + H)^+$ at m/z 149 and $(M - 2NO_2 - CH_2NNO_2)^+$ at m/z 130 do not appear in the FD spectrum of RDX.

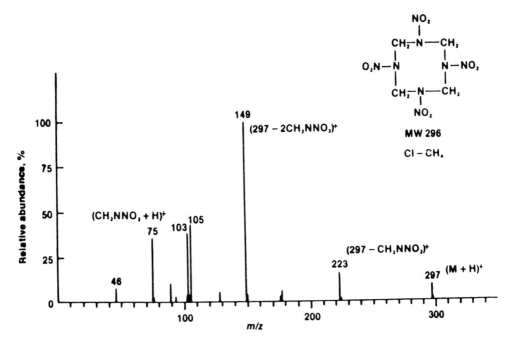

FIGURE 8. CI-methane mass spectrum of HMX. (From Vouros, P. et al., *Anal. Chem.*, 49, 1309, 1977. With permission.)

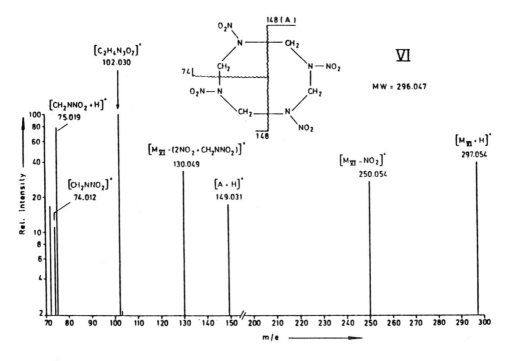

FIGURE 9. FD mass spectrum of HMX. (From Schulten, H.-R. and Lehmann, W. D., *Anal. Chim. Acta*, 93, 19, 1977. With permission.)

D. Various Explosives

Two organic peroxides which were used in terrorist activities[37] are triacetone-triperoxide (TATP, 2) and hexamethylenetriperoxidediamine (HMTD, 3).

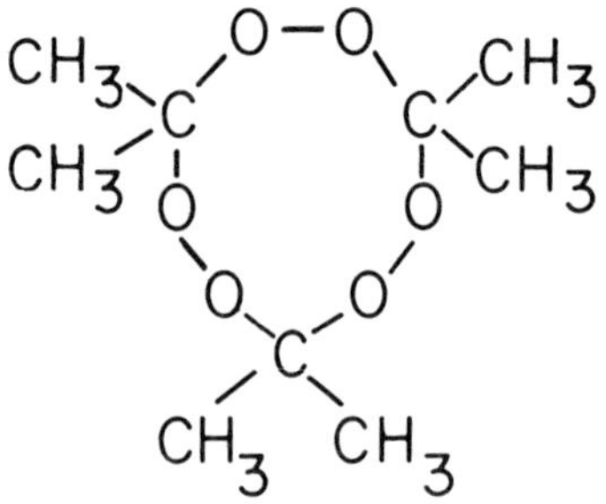

2

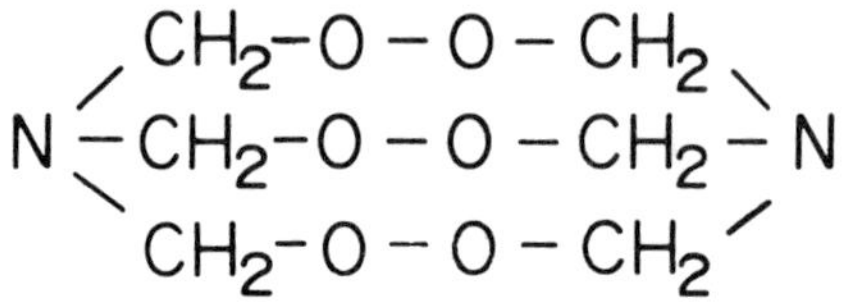

3

The EI mass spectrum of TATP[37] contains ions only in the low mass range with a base peak at m/z 43.

The CI-isobutane mass spectrum of TATP,[37] shown in Figure 10, contains an MH+ ion at m/z 223.

The EI mass spectrum of HMTD[33,37,38] has its most abundant ions in the low mass range: at m/z 30, 42, 45, 58, 73, and 88. It contains also ions at m/z 176 and m/z 208 (M+).

The CI mass spectra of HMTD were recorded with isobutane,[33,37] with methane,[33,38] and with ammonia[33] as reagents. The CI-isobutane spectrum of HMTD[37] is shown in Figure 11.

E. Mixtures

Commercial dynamites, military explosives, and smokeless powders are usually produced as mixtures containing both explosive and nonexplosive ingredients. It is therefore of great interest to be able to identify such explosives by MS.

Most nonexplosive additives are stabilizers and plasticizers. Table 2 gives a list of the CI mass spectral ions of some of these additives.[14]

Although EI does not seem to be the ideal method for the analysis of mixtures, some results of analysis of mixtures by EI were obtained. The EI mass spectrum of pentolite 50/50 (a mixture of TNT and PETN)[39] was found to be similar to the spectrum of TNT with one exception: the peak at m/z 46, which is the base peak in the EI spectrum of PETN, was much larger in the mass spectrum of pentolite than in TNT.

Composition B contains 58.2% RDX, 40% TNT, 1.2% polyisobutylene, and 0.6% wax. The EI mass spectrum of composition B at 60°C was similar to that of TNT, and at 180°C it was similar to that of RDX.[39]

Composition C-4 contains 91% RDX, 2.1% polyisobutylene, 1.6% motor oil, and 5.3% di(2-ethylhexyl) sebacate, a plasticizer. The EI mass spectrum of C-4 was found to be similar to that of RDX.[39]

The CI mass spectra of an explosive mixture used in letter bombs have been recorded with methane[4] and water[30] as reagents. The CI-water mass spectrum of this explosive

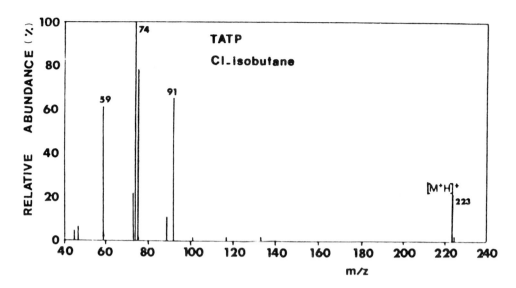

FIGURE 10.   CI-isobutane mass spectrum of TATP. (From Zitrin, S. et al., *Proc. Int. Symp. Analysis and Detection of Explosives,* FBI Academy, Quantico, Va., 1983, 137. With permission.)

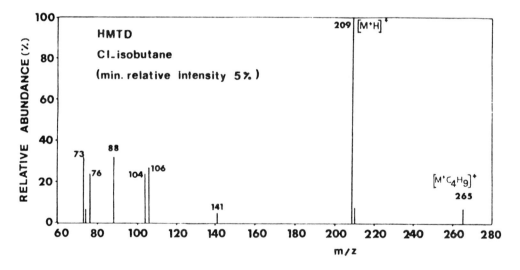

FIGURE 11.   CI-isobutane mass spectrum of HMTD. (From Zitrin, S. et al., *Proc. Int. Symp. Analysis and Detection of Explosives,* FBI Academy, Quantico, Va., 1983, 137. With permission.)

mixture is shown in Figure 12. The mass spectrum consists of the CI mass spectra of both RDX and PETN and a combination of both. A typical ion is the $(M_{RDX} + M_{PETN} + H)^+$ ion at m/z 539.

FD has been used also for the identification of explosive mixtures. The FD mass spectrum of a mixture of RDX and 10% HMX[22] produced the M^+ and MH^+ ions of HMX at m/z 296 and 297, respectively, and the M^+ and MH^+ ions of RDX at m/z 222 and m/z 223, respectively, as well as a series of common fragment ions. The FD mass spectrum of pure HMX did not produce any ions at m/z 222 or 223.

The FD mass spectrum of a technical explosive mixture[22] containing 39.5% TNT, 59.5% RDX, and 1% wax is shown in Figure 13. The mass spectrum includes the M^+ and MH^+ ions of TNT at m/z 227 and 228, respectively, and the MH^+ ion of RDX at

Table 2
CI-METHANE MASS SPECTRAL IONS OF SOME ADDITIVES[12]

Compound	Molecular weight (amu)	Ion — m/z (abundance, %)							
		$C_6H_5O^+_3$	$(MH - C_6H_5NHC_2H_5 - CO)^+$	$(MH - C_6H_5NHC_2H_5)^+$	$(MH - C_4H_{10}O)^+$	$(MH - NO_3)^+$	MH^+	$(M + C_2H_5)^+$	$(M + C_3H_5)^+$
Diphenylamine	169						170 (100)	198 (16)	210 (1.3)
2-Nitro-diphenylamine	214					169 (1.6)	215 (100)	243 (8)	255 (2.4)
Methyl centralite	240			134 (13)			241 (100)	269 (10)	271 (2)
Ethyl centralite	268		120 (3)	148 (25)			269 (100)	297 (20)	309 (2.5)
Dibutyl phthalate	278	149 (100)			205 (65)		279 (22)	307 (3)	

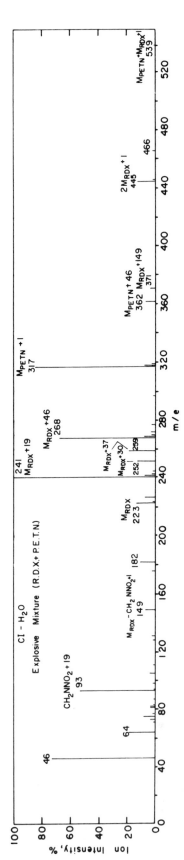

FIGURE 12.   CI-water mass spectrum of a letter bomb explosive. (From Yinon, J., *Biom. Mass Spectrom.*, 1, 393, 1974. With permission.)

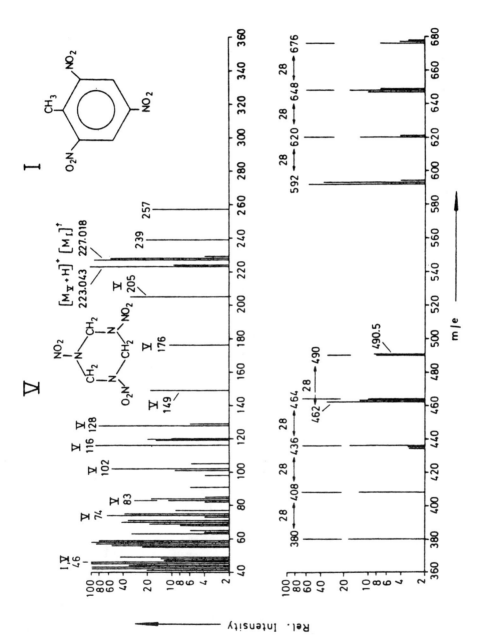

FIGURE 13. FD mass spectrum of a technical explosive containing TNT, RDX, and wax. (From Schulten, H.-R. and Lehmann, W. D., *Anal. Chim. Acta*, 93, 19, 1977. With permission.)

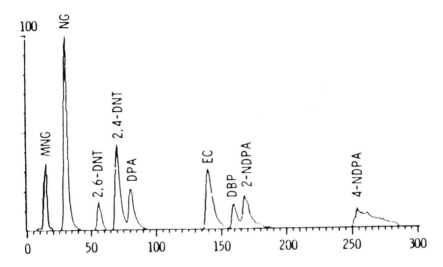

FIGURE 14.   Total ion chromatogram of smokeless powder from a 0.22-caliber cartridge.
(From Mach, M. H. et al., *J. Forensic Sci.,* 23, 433, 1978. With permission.)

m/z 223. The wax, which is composed of a number of fatty acids, produces a series of
ions differing by 28 amu due to the C_2H_4 groups of these fatty acids.

MS has also been used in the forensic laboratory to differentiate between single-base
and double-base smokeless gunpowders.[40] Nitrocellulose (NC) is the basic ingredient
in single-base gunpowder, while NG, added to NC in quantities varying from 1 to 40%,
is the characteristic ingredient of double-base gunpowders. A series of 18-canister
smokeless gunpowders were examined by MS headspace analysis using EI and CI-
methane. Double-base gunpowders will produce mass spectra which include typical
ions of NG like the MH$^+$ ion at m/z 228 and (MH − HNO$_3$)$^+$ at m/z 165, while mass
spectra of single-base gunpowders will not include these ions.

GC/MS is one of the most widely used methods for the analysis of complex mixtures. It has, however, been used to only a limited extent for the analysis of explosives
for reasons pointed out previously.

GC/MS has been used for the analysis of smokeless gunpowders. The compositions
of 32 commercial smokeless powders were determined[41] using GC/MS with EI and CI-
methane. The gas chromatogram of smokeless powder from a 5.6-mm (0.22-caliber)
Western Super-X 40 grain Lubaloy RNL Cartridge[41] is shown in Figure 14. Five major
components were found: NG, 2,4,-DNT, diphenylamine (DPA), dibutyl phthalate
(DBP), and ethyl centralite (EC, 1,3-diethyl-1,3-diphenylurea). The various components were positively identified by EI and CI-methane MS.

A GC/MS system has been used for the study of impurities in crude TNT samples.[42]
The identified impurities included *p*- and *m*-DNB, all the DNT isomers (except 3,5-
DNT), TNB, TNX, 2,4,5-, and 2,3,4-TNT and 1-nitronaphthalene. The GC-separated
impurities were identified by a high-resolution mass spectrometer.

2,4-DNT, 2,4,6-TNT, and two reduction products of TNT — 2-amino-4,6-dinitro-
toluene and 4-amino-2,6-dinitrotoluene — have been identified in groundwater contaminated with TNT wastewater, using a GC/MS system.[43]

GC/MS has also been used in combination with CI. Figure 15 shows the total ion
chromatogram of Hi-Drive® dynamite with an added internal standard (dimethyl is-
ophthalate).[14] The CI-methane mass spectra of the two explosive ingredients, EGDN
and NG, are shown in Figure 16.[14] Both compounds produce MH$^+$ ions, adduct ions
(M + NO)$^+$, and (M + NO$_2$)$^+$ and undergo a similar fragmentation process leading to
the formation of (MH − HNO$_3$)$^+$.

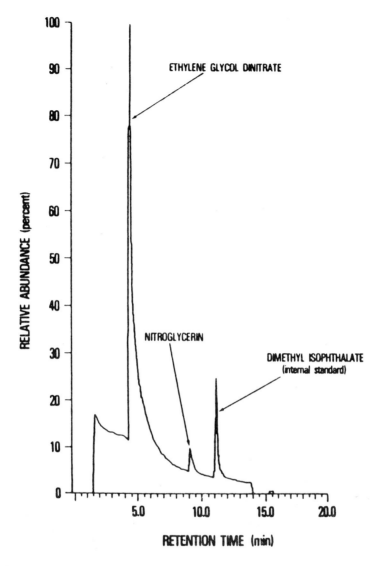

FIGURE 15. Total ion chromatogram of Hi-Drive® dynamite. (From Pate, C. T. and Mach, M. H., *Int. J. Mass Spectrom. Ion Phys.*, 26, 267, 1978. With permission.)

Residue from a terrorist bombing was analyzed by GC/MS using a fused silica capillary column and CI-methane ionization.[38] The CI mass spectra of the two obtained GC peaks indicated that the components were DEGN and metriol trinitrate (MTN), two components of a U.S.-made dynamite. The CI-methane mass spectrum of MTN[38] contained an abundant MH+ ion at m/z 256 and a series of fragment ions, the most abundant ones being at m/z 146 and m/z 193 (base peak).

Pyrolysis/MS has been used to detect and identify undecomposed explosive traces after detonation.[44] Post-detonation carbon residues were taken from cloth and from wood and analyzed by pyrolysis/MS using EI ionization with single ion monitoring of the (M − OH)+ ion of TNT at m/z 210. TNT was found in 20% of the samples examined.

HPLC is a good separation technique for both volatile and nonvolatile compounds and is usually done with the column at room temperature. It is therefore a suitable method for separation of thermally labile compounds in complex mixtures. The com-

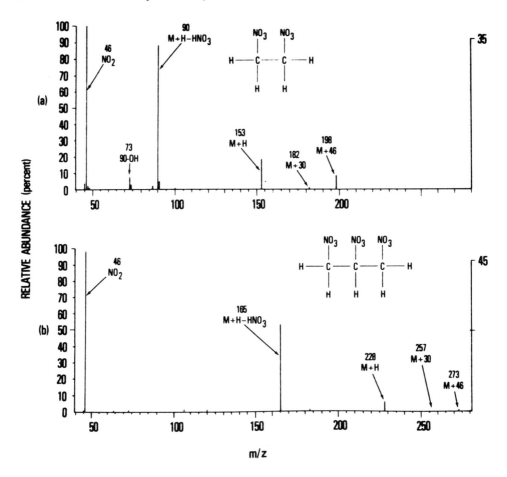

FIGURE 16. CI-methane mass spectra of EGDN and NG. (From Pate, C. T. and Mach, M. H., *Int. J. Mass Spectrom. Ion Phys.*, 26, 267, 1978. With permission.)

bination of the chromatographic separation obtained by HPLC with the sensitivity and specificity of MS constitutes a powerful technique for the analysis of explosives.

Two on-line liquid chromatograph/mass spectrometer (LC/MS) systems have been used successfully for the analysis of explosive mixtures.[45,46] One LC/MS system[45] uses a quadrupole mass spectrometer with a direct liquid introduction LC/MS interface. NCI was chosen as the ionization mode, because in this particular system it was found to produce more abundant ions in the investigated explosives. An example which demonstrates the use of this LC/MS system is shown in Figures 17 and 18. Figure 17 shows the HPLC chromatogram of a military explosive, tetrytol, as obtained on the UV detector and as a total ion current (TIC) trace. An RP-18 reversed-phase column was used with methanol-water (50:50) as mobile phase. About 1% of the liquid effluent enters the mass spectrometer ion source where the mobile phase serves as NCI moderating reagent. Figure 18 shows the LC/MS NCI mass spectra of the two explosive components of tetrytol, tetryl, and TNT.

Another LC/MS system used for the analysis of explosives[46] is based on a magnetic sector mass spectrometer with a direct liquid introduction LC/MS interface. An example is shown in Figures 19 to 21. Figure 19 shows the HPLC/UV trace of a technical mixture containing TNT and RDX. An RP-8 reversed-phase column was used with acetonitrile-water (50:50) as mobile phase. The ionizing technique used in this system was CI, the mobile phase serving as reagent. Figures 20 and 21 show the LC/MS mass spectra of the two components, TNT and RDX.

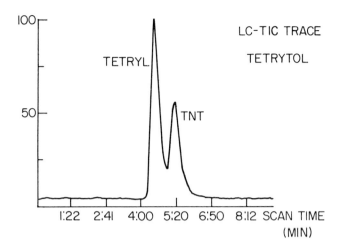

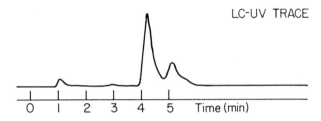

FIGURE 17. HPLC chromatogram of tetrytol (TIC and UV traces). (From Parker, C. E. et al., *J. Forensic Sci.*, 27, 495, 1982. With permission.)

III. DETECTION OF HIDDEN EXPLOSIVES

The detection of hidden explosives has become a highly important problem in forensic science. With the increasing use of explosives by terrorist groups and by individuals, law enforcement agencies throughout the world are faced with the problem of detecting these explosives in suitcases, mail, vehicles, aircraft, etc.

The main requirements of an explosive detector are sensitivity and specificity. Additional desirable characteristics are simplicity, reliability, and fast response.

A major problem, when using a mass spectrometer as an explosive vapor detector, is the introduction of the ambient air, to be checked for explosive vapor, into the ion source which is under vacuum.

Several mass spectrometers were developed for explosives detection. Only a few examples will be presented here. The Universal Monitor Corporation Olfax explosive detector[47,48] was based on a quadrupole mass spectrometer with an EI source. Ambient air was introduced through a dual-membrane inlet system consisting of two dimethyl silicone membranes. Sensitivity to TNT was estimated to be 1 ppb, monitoring the $(M - OH)^+$ ion at m/z 210.

A portable EI quadrupole mass spectrometer for explosive vapor detection, housed in two aluminum suitcases, was developed by Varian Associates.[49] The inlet system consisted of a three-stage membrane separator. Since the dimethyl silicone membrane is more permeable to organic molecules than to the air gases, the relative concentration of the transmitted explosive is increased at each stage of the separator. Sample enrichment was about 10^6. It was found that TNT was best transmitted at a separator temperature of 200°C. The more volatile mononitrotoluene (MNT) and dinitrotoluene

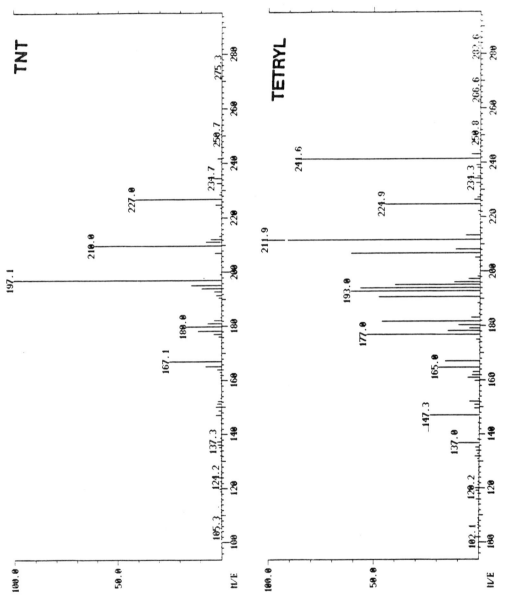

FIGURE 18.   LC/MS mass spectra of tetryl and TNT. (From Parker, C. E. et al., *J. Forensic Sci., 27*, 495, 1982. With permission.)

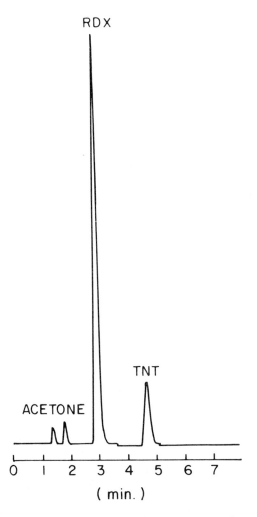

FIGURE 19. HPLC/UV trace of a technical mixture containing TNT and RDX. (From Yinon, J. and Hwang, D.-G., *J. Chromatogr.*, 268, 45, 1983. With permission.)

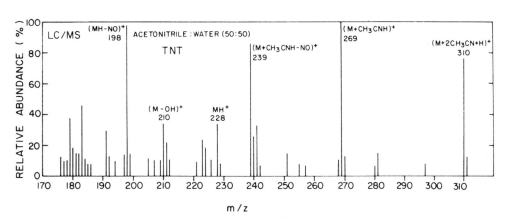

FIGURE 20. LC/MS mass spectrum of TNT. (From Yinon, J. and Hwang, D.-G., *J. Chromatogr.*, 268, 45, 1983. With permission.)

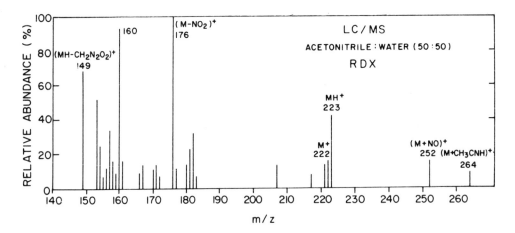

FIGURE 21. LC/MS CI mass spectrum of RDX. (From Yinon, J. and Hwang, D.-G., *J. Chromatogr.*, 268, 45, 1983. With permission.)

(DNT), present as impurities in technical grade TNT, were detected more easily than TNT. The detection sensitivity for MNT was found to be 2 ppb.

The TAGA 3000 system explosive detector developed by Sciex Inc.[50] and mounted in a mobile van, is based on positive- and negative ion API/MS. Ambient air is drawn into the inlet at flow rates up to 10 *l*/sec. Reactant ions are produced in the source by corona discharge. Sample and reactant ions are focused onto an aperture and transferred into a quadrupole mass analyzer through a series of ion lenses. The detection limit for TNT, monitoring the $(M - H)^-$ ion at m/z 226 was found to be less than 1 ppt.[51] Experiments with the TAGA system were done[51] by taking ambient carrier gas from the cabin exhaust of an aircraft. NG was detected at a concentration of a few parts per trillion in real-time by its mass spectrum, when dynamite was present in the aircraft cabin. The detection was based on the API mass spectrum of NG characterized by the ions $[(M_{NG} - H - NO_2) + O_2]^-$ at m/z 212 and $[(M_{NG} - NO_2) + O_2]^-$ at m/z 213.

One of the most significant recent advances in MS is the development of tandem mass spectrometry (MS/MS).[52,53] MS/MS has already been used for the study of fragmentation processes in RDX and HMX.[36]

Initial experiments with a MS/MS as an explosive detector were done[54] using the TAGA 6000 MS/MS, developed by Sciex Inc. This MS/MS system consists of a triple quadrupole mass spectrometer with an API source operating in either the positive or negative ion mode.

The main features of MS/MS are specificity, sensitivity, and fast response. Therefore, its future applications[55] in residue analysis and detection of explosives seem obvious.

REFERENCES

1. Yinon, J., Mass spectrometry of explosives: nitro compounds, nitrate esters, and nitramines, *Mass Spectrom. Rev.*, 1, 257, 1982.
2. Volk, F. and Schubert, H., Massenspektrometrische Untersuchungen von Explosivstoffen, *Explosivstoffe*, 16, 2, 1968.

3. Murrman, R. P., Jenkins, T. F., and Leggett, D. C., Composition and mass spectra of impurities in military grade TNT vapor, *USA CRREL Special Report 158,* Cold Regions Research and Engineering Laboratory, Hanover, N.H., 1971.

4. Zitrin, S. and Yinon, J., Chemical ionization mass spectrometry of explosives, in *Advances in Mass Spectrometry in Biochemistry and Medicine,* Vol. 1, Frigerio, A. and Castagnoli, N., Eds., Spectrum, New York, 1976, 369.

5. Alm, A., Dalman, O., Frölen-Lindgren, I., Hulten, F., Karlsson, T., and Kowalska, M., *Analysis of Explosives,* FOA Report C 20267-D1, National Defense Research Institute, S-104 50 Stockholm, Sweden, 1978.

6. Zitrin, S. and Yinon, J., Mass spectrometry studies of trinitroaromatic compounds, in *Advances in Mass Spectrometry,* Vol. 7B, Daly, N. R., Ed., Heyden & Son, London, 1978, 1457.

7. Bulusu, S. and Axenrod, T., Electron impact fragmentation mechanisms of 2,4,6-trinitrotoluene derived from metastable transitions and isotope labeling, *Org. Mass Spectrom.,* 14, 585, 1979.

8. Dorey, R. C. and Carper, W. R., Synthesis and high-resolution mass-spectral analysis of isotopically labeled 2,4,6-trinitrotoluene, *J. Chem. Eng. Data,* 29, 93, 1984.

9. Meyerson, S., Puskas, I., and Fields, E. K., Organic ions in the gas phase. XVIII. Mass spectra of nitroarenes, *J. Am. Chem. Soc.,* 88, 4974, 1966.

10. Cumming, A. S. and Park, K. P., The analysis of trace levels of explosive by gas chromatography-mass spectrometry, in *Proc. Int. Symp. Analysis and Detection of Explosives,* FBI Academy, Quantico, Va., 1983, 259.

11. Gillis, R. G., Lacey, M. J., and Shannon, J. S., Chemical ionization mass spectra of explosives, *Org. Mass Spectrom.,* 9, 359, 1974.

12. Saferstein, R., Chao, J. M., and Manura, J. J., Isobutane chemical ionization mass spectrographic examination of explosives, *J. Assoc. Off. Anal. Chem.,* 58, 734, 1975.

13. Vouros, P., Peterson, B. A., Colwell, L., Karger, B. L., and Harris, H., Analysis of explosives by high performance liquid chromatography and chemical ionization mass spectrometry, *Anal. Chem.,* 49, 1309, 1977.

14. Pate, C. T. and Mach, M. H., Analysis of explosives using chemical ionization mass spectroscopy, *Int. J. Mass Spectrom. Ion Phys.,* 26, 267, 1978.

15. Yinon, J. and Laschever, M., Reduction of trinitroaromatic compounds in water by chemical ionization mass spectrometry, *Org. Mass Spectrom.,* 16, 264, 1981.

16. Yinon, J. and Laschever, M., Direct-injection chemical ionization mass spectrometry of explosives in water, *Eur. J. Mass Spectrom. Biochem. Med. Environ. Res.,* 2, 101, 1982.

17. Yinon, J. and Zitrin, S., Processing and interpreting mass spectral data in forensic identification of drugs and explosives, *J. Forensic Sci.,* 22, 742, 1977.

18. Yinon, J., Direct exposure chemical ionization mass spectra of explosives, *Org. Mass Spectrom.,* 15, 637, 1980.

19. Yinon, J., Boettger, H. G., and Weber, W. P., Negative ion mass spectrometry — a new analytical method for detection of trinitrotoluene, *Anal. Chem.,* 44, 2235, 1972.

20. Yinon, J., Analysis of explosives by negative ion chemical ionization mass spectrometry, *J. Forensic Sci.,* 25, 401, 1980.

21. Mitchum, R. K., Althaus, J. R., Korfmacher, W. A., and Moler, G. F., Application of negative ion atmospheric pressure ionization (NIAPI) mass spectrometry for trace analysis, in *Advances in Mass Spectrometry,* Vol. 8B, Quayle, A., Ed., Heyden & Son, London, 1980, 1415.

22. Schulten, H.-R. and Lehmann, W. D., High-resolution field desorption mass spectrometry. VII. Explosives and explosive mixtures, *Anal. Chim. Acta,* 93, 19, 1977.

23. Jenkins, T. F., Murrmann, R. P., and Leggett, D. C., Mass spectra of isomers of trinitrotoluene, *J. Chem. Eng. Data,* 18, 438, 1973.

24. Deutsch, J. and Sklarz, B., Mass spectra of amine picrates, *Isr. J. Chem.,* 10, 51, 1972.

25. Zitrin, S. and Yinon, J., Chemical ionization mass spectra of 2,4,6-trinitroaromatic compounds, *Org. Mass Spectrom.,* 11, 388, 1976.

26. Yinon, J., Atmospheric pressure ionization (API) mass spectrometry of explosives, presented at 26th Annual Conference on Mass Spectrometry and Allied Topics, St. Louis, Mo., 1978. Book of Abstracts, p. 118.

27. Yelton, R. O., Negative (CCl₄) and positive (NH₃) chemical ionization of the explosive hexanitrostilbene, presented at 30th Annual Conference on Mass Spectrometry and Allied Topics, Honolulu, Hawaii, 1982. Book of Abstracts, p. 665.

28. Fraser, R. T. M. and Paul, N. C., The mass spectrometry of nitrate esters and related compounds. I, *J. Chem. Soc. (B),* 659, 1968.

29. Yinon, J. and Laschever, M., unpublished data, 1981.

30. Yinon, J., Identification of explosives by chemical ionization mass spectrometry using water as reagent, *Biomed. Mass Spectrom.,* 1, 393, 1974.

31. Yelton, R. O., Ammonia chemical ionization of the explosives pentaerythritoltetranitrate and dipentaerythritol hexanitrate, presented at 30th Annual Conference on Mass Spectrometry and Allied Topics, Hololulu, Hawaii, 1982. Book of Abstracts, p. 667.

32. St. John, G. A., McReynolds, J. H., Blucher, W. G., Scott, A. C., and Anbar, M., Determination of the concentration of explosives in air by isotope dilution analysis, *Forensic Sci.,* 6, 53, 1975.

33. Gielsdorf, W., Identifizierung einiger Sprengstoffe mit Hilfe spezieller GC/MS-Techniken, inbesondere der PPNICI-Methode, *Fresenius Z. Anal. Chem.,* 308, 123, 1981.

34. Bulusu, S., Axenrod, T., and Milne, G. W. A., Electron-impact fragmentation of some secondary aliphatic nitramines. Migration of the nitro group in heterocyclic nitramines, *Org. Mass Spectrom.,* 3, 13, 1970.

35. Stals, J., Chemistry of aliphatic unconjugated nitramines. VII. Interrelations between the thermal, photochemical and mass spectral fragmentation of RDX, *Trans. Faraday Soc.,* 67, 1768, 1971.

36. Yinon, J., Harvan, D. J., and Hass, J. R., Mass spectral fragmentation pathways in RDX and HMX. A mass analyzed ion kinetic energy spectrometric/collisional induced dissociation study, *Org. Mass Spectrom.,* 17, 321, 1982.

37. Zitrin, S., Kraus, S., and Glattstein, B., Identification of two rare explosives, in *Proc. Int. Symp. Analysis and Detection of Explosives,* FBI Academy, Quantico, Va., 1983, 137.

38. Reutter, D. J., Bender, E. C., and Rudolph, T. L., Analysis of an unusual explosive: methods used and conclusions drawn from two cases, in *Proc. Int. Symp. Analysis and Detection of Explosives,* FBI Academy, Quantico, Va., 1983, 149.

39. Coates, A. D., Freedman, E., and Kuhn, L. P., Characteristics of certain military explosives, Report No. 1507, Ballistic Research Laboratory, Aberdeen Proving Ground, Maryland, 1970.

40. Hardy, D. R. and Chera, J. J., Differentiation between single-base and double-base gunpowders, *J. Forensic Sci.,* 24, 618, 1979.

41. Mach, M. H., Pallos, A., and Jones, P. F., Feasibility of gunshot residue detection via its organic constituents. I. Analysis of smokeless powders by combined gas chromatography-chemical ionization mass spectrometry, *J. Forensic Sci.,* 23, 433, 1978.

42. Chang, T. L., Identification of impurities in crude TNT by tandem GC-MS technique, *Anal. Chim. Acta,* 53, 445, 1971.

43. Pereira, W. E., Short, D. L., Manigold, D. B., and Roscio, P. K., Isolation and characterization of TNT and its metabolites in groundwater by gas chromatograph-mass spectrometer — computer techniques, *Bull. Environ. Contam. Toxicol.,* 21, 554, 1979.

44. Cumming, A. S., Park, K. P., and Clench, M. R., The analysis of post-detonation carbon residues by mass spectrometry, in *Proc. Int. Symp. Analysis and Detection of Explosives,* FBI Academy, Quantico, Va., 1983, 235.

45. Parker, C. E., Voyksner, R. D., Tondeur, Y., Henion, J. D., Harvan, D. J., Hass, J. R., and Yinon, J., Analysis of explosives by liquid chromatography — negative ion chemical ionization mass spectrometry, *J. Forensic Sci.,* 27, 495, 1982.

46. Yinon, J., and Hwang, D.-G., High-performance liquid chromatography-mass spectrometry of explosives, *J. Chromatogr.,* 268, 45, 1983.

47. Spangler, G. E., Membrane technology in trace gas detection. I. Evaluation of the Universal Monitor Olfax instrument, Report 2083, U.S. Army Mobility Equipment Research and Development Center, Fort Belvoir, Va., 1973.

48. Spangler, G. E., Analysis of two membrane inlet systems on two potential trace vapor detectors, *Int. Lab.,* 24, July/August, 1975.

49. Evans, J. E. and Arnold, J. T., Monitoring organic vapors, *Environ. Sci. Technol.,* 9, 1134, 1975.

50. Reid, N. M., Buckley, J. A., Lane, D. A., Lovett, A. M., French, J. B., and Poon, C., Taga 3000 — a new digitally controlled APCI mass spectrometer based system, presented at 26th Annual Conference on Mass Spectrometry and Allied Topics, St. Louis, Missouri, 1978. Book of Abstracts, p. 662.

51. Buckley, J. A., French, J. B., and Reid, N. M., Taga — a total system approach to sub-ppt explosive detection by natural vapour identification, in Proc. New Concepts Symposium and Workshop on the Detection and Identification of Explosives, Reston, Virginia, 1978, 109.

52. McLafferty, F. W., Tandem mass spectrometry, *Science,* 214, 280, 1981.

53. Yost, R. A. and Fetterolf, D. D., Tandem mass spectrometry (MS/MS) instrumentation, *Mass Spectrom. Rev.,* 2, 1, 1983.

54. Tanner, S. D., Davidson, W. R., and Fulford, J. E., The instantaneous detection of explosives by tandem mass spectrometry, in *Proc. Int. Symp. Analysis and Detection of Explosives,* FBI Academy, Quantico, Va., 1983, 409.

55. Yinon, J., New horizons in mass spectrometry for the analysis of explosives, in *Proc. Int. Symp. Analysis and Detection of Explosives,* FBI Academy, Quantico, Va., 1983, 1.

Chapter 5

ARSON ANALYSIS BY MASS SPECTROMETRY

R. Martin Smith

TABLE OF CONTENTS

I. BACKGROUND

A. Introduction

It seems ironic that mass spectrometry (MS) which, during its developmental years, was an analytical workhorse for the petroleum industry, is only now beginning to find routine application in the analysis of the primarily petroleum-derived residues of arson fires. This irony is intensified because combining the high sensitivity and specificity of MS with the tremendous separation power of gas chromatography (GC/MS) should provide the ideal technique for examining these complex residues.

Yet there are several legitimate reasons for this gap. Until recently, many forensic laboratories (both privately and governmentally funded) lacked either the money and/ or qualified personnel to maintain mass spectrometers at their facilities; even those laboratories with the instrumentation often were frustrated because the extreme complexity of many samples simply overwhelmed the data-handling capabilities of the unit. In our own case, identifying up to a score or more components per sample via manually collected and counted spectra was an odious, and not necessarily productive, task. Perhaps most importantly at this time, however, is the fact that even the most sophisticated GC/MS/data system (GC/MS/DS) cannot assure the answer to a crucial question: what relation, if any, does the identification of certain volatile organic chemicals in the debris from a fire have to any accelerant which may have been used to start the fire? It is the answer to this question, and not the identification of the individual compounds themselves, that demands the expertise of the arson analyst.

B. History

It is difficult to talk about the MS of arson residues without talking about combined GC/MS. Most residues encountered in this field are chemically too complex to be treated without some form of separation, despite recent attempts using positive chemical ionization mass spectrometry (CIMS) to characterize intact complex hydrocarbon mixtures.[1] In fact, GC used alone is still the essential feature of current arson analyses.[2-4] Even high-pressure liquid chromatography (HPLC), which has become popular in many other areas of chemical analysis, has made few inroads into the analysis of petroleum-related compounds.[5-6]

Although Zoro and Hadley reported the routine use of GC/MS for arson cases in Birmingham, U.K. as early as 1976,[7] its use as a primary identification tool in arson analysis has evolved rather slowly. Even as late as 1982 a chapter on arson investigation in an extensive book on forensic science devoted less than one paragraph to GC/MS.[8] The earliest published applications of this technique to arson work dealt mainly with samples or situations of fairly limited scope. Our own work, for example, was severely limited for several years by low sensitivity, poor chromatographic resolution, and totally inadequate data-handling capabilities. One example,[9] showing the presence of chloroform, toluene, and amyl alcohol in a rag from a factory fire, was fairly novel at the time, but rather unexciting by current standards. The recovery by Maucieri and Bowden of piperidine, cyclohexanone, and bromobenzene from the remains of a simulated phencyclidine ("PCP") drug lab fire provided a highly specialized application with a similar lack of sample complexity.[10] More ambitious was Mach's attempt to characterize gasoline residues by GC/MS,[11] where primary emphasis was placed upon characterization of the high-boiling aromatics using methane CIMS.

Recent applications have revolved around the use of mass chromatography to locate compounds in complex samples by screening for certain characteristic ions.[12-16] This technique offers a productive approach to arson analysis in general and will be discussed more fully below.

C. Sample Considerations

The handling and preliminary treatment of arson samples often is not straightforward. Sample collection and delivery to the laboratory is particularly critical. Investigators must obtain appropriate questioned and standard samples within a relatively short period of time after the fire to ensure retention of any volatiles present. Questioned samples hopefully contain materials still impregnated with accelerant from the origin of the fire, whereas standard samples must be collected far enough away from the source of the fire to preclude the presence of volatile accelerants, yet close enough to provide representative samples of construction and furnishing materials for comparison of pyrolysis products. These samples are then individually sealed in unreactive and impervious containers such as unused metal paint cans, glass canning jars,[2] or even polyester bags.[17]

Older methods of isolating volatiles from debris samples, such as steam distillation and solvent extraction, are rapidly being replaced by those using adsorption onto activated charcoal or similar material. In one method ("purge-and-trap") inert gas (usually nitrogen) is passed over the debris in a heated can, then through a trap of activated charcoal[18] or porous material such as Tenax®.[19] Trapped compounds are then either eluted with a noninterfering solvent such as carbon disulfide or thermally desorbed directly into the injection port of a gas chromatograph.[19-20] A more recent method involves placing charcoal-covered beads[21] or Curie-point wires[22] directly above the debris for short periods of time, also followed by thermal desorption into the gas chromatograph. Both these methods appear to be even more sensitive than the more direct and conventional headspace analysis, which also suffers by being prejudiced against higher-boiling components.[20]

Although packed columns are still widely used for GC work in arson analysis, a significant trend has begun towards the use of capillary columns.[23,24] The new flexible and durable fused silica capillary columns seem ideally suited for this type of work.[25,26] Although a wide variety of hydrocarbon-specific column coatings have been used, methyl silicone coatings (which are thermally more stable and generally produce less column bleed) and temperature programs of roughly 50 to 250°C over 30 to 60 min appear adequate for much GC/MS work.

Nearly all mass spectrometric analyses of arson materials have been done using electron impact (EI) ionization. Although both positive and negative chemical ionization (PCI, NCI) have utility in the characterization of aromatic compounds,[1,27] they seem to offer no real advantages in either sensitivity or specificity for the identification of either the aromatic or aliphatic compounds normally encountered in arson residues. The remaining discussions will assume positive ion electron impact ionization conditions unless otherwise noted.

II. CHARACTERIZATION TECHNIQUES

A. What Constitutes Identification?

The characterization of an accelerant from an arson residue as paint thinner, evaporated gasoline, or fuel oil is not the same as the identification of the same materials fresh from a can. In the latter case, comparison of gas chromatographic patterns (retention times plus relative concentrations) with those of appropriate standards followed by identification of at least the major constituents by their mass spectra would usually suffice to identify the material to a high degree of certainty. However, the intense heat and oxidizing conditions of a typical arson fire significantly alters the chemical make-up of any accelerant present except in unusual circumstances. Evaporation, oxidation, pyrolysis, and contamination by pyrolysis products of structural materials are all real and serious problems which hinder identification of the original accelerant (Figure 1).[28] The analyst thus must use his or her experience to piece together what information is available to come up with a "characterization"[29] of the material, even if its exact nature cannot be specified.

How much information is necessary to identify a specific accelerant or even characterize a general accelerant classification is somewhat problematic. Historically, gasoline and fuel oil have been almost the only accelerants identifiable per se from arson residues, partly because they appear to be the most popular accelerants[30] (hence they occur more often and more analysts are familiar with them), but also because they both have a number of high-boiling components in characteristic proportions which have proven to be almost impervious to fire conditions. Their identification is often easy even by GC alone, most authors agreeing that several low- to medium-boiling aromatics best characterize gasoline, while a strong series of n-alkanes in the C_{10} to C_{18} range indicates fuel oil.[11,31,32]

With other accelerants the situation is more complex. Currently published standards, which permit GC characterization of "medium petroleum distillates" on the basis of "a pattern starting between C_8 and C_{10}, ending near C_{12} and containing at least three major peaks between C_8 and C_{12},"[32] scarcely seem sufficient in view of the seriousness of the issue. Instead, it would seem that characterization of any accelerant should be based on comparison of the composition of the arson residue with the known compositions of many accelerants, their evaporated and weathered residues, and possible contaminating pyrolysis products. This process is best accomplished by GC/MS since, as will be seen later on, the presence of a suspicious GC pattern does not ensure the presence of an accelerant.

B. Standard Mass Spectra

Most of the organic compounds encountered in arson accelerants are hydrocarbons, since they are easily the predominant constituents of commercial petroleum products.[33,34] Oxygenated compounds are much less frequent, limited mainly to specific solvent mixtures such as lacquer thinners[34] or as preservatives,[13] while nitrogenous and halogenated compounds are even rarer. The possibilities are enormous, however, as the following list of contents of an undoubtedly flammable nail polish remover will attest: acetone, mineral spirits (medium-volatility hydrocarbons), isopropanol, ethyl acetate, mineral oil, cottonseed oil, lanolin oil, triethanolamine, and terpineol.

Mass spectral identification of purified organic compounds demands a collection of standard spectra for comparison, either run on the user's own instrument or contained in a library collection of known quality. Although the first choice undoubtedly minimizes the problems of interinstrument variations in spectral quality,[35] it is somewhat impractical for arson work, since at least several hundred compounds (many commercially unavailable or very expensive in pure form) are necessary. On the other hand, large validated mass spectral libraries are available today in book form,[36] or more conveniently are included on magnetic tape or disc for computer searching with most commercially available GC/MS/DS units.

Even if the quality of library spectra is uniformly good, identification of arson residue mass spectra by computer library search often is less than ideal.[13] In our own work, some compounds, most notably the n-alkanes, were consistently misidentified by the search program. Others, such as the isomeric alkylbenzenes, gave virtually indistinguishable spectra, making individual compound identification extremely difficult. Improving GC column resolution by switching from packed to capillary columns simplified background subtraction somewhat, but did not alleviate the problem. Indeed, it became apparent that GC resolution would never be sufficient to adequately separate

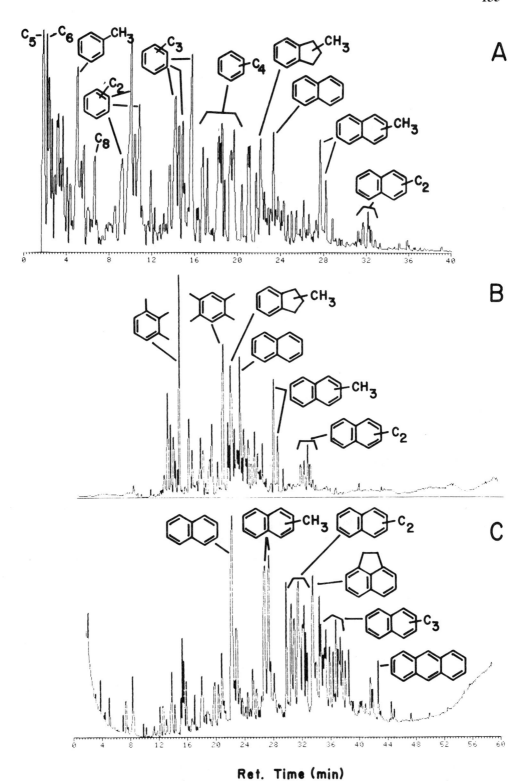

FIGURE 1. Chromatograms of gasoline at various stages of evaporation: (A) raw gasoline; (B) evaporated gasoline; (C) highly evaporated and weathered gasoline. (From Smith, R. M., *Anal. Chem.*, 54, 1399A, 1982. With permission.)

all of the components of the most complex mixtures,[37,38] and that instrumental differences would preclude absolute identification of the mass spectra of many compounds.

C. Mass Chromatography

Mass spectral examination of complex arson mixtures thus must be approached by some sort of compromise between complete resignation and unequivocal compound identification. This seems best achieved by mass or reconstructed ion chromatography, first described by Biemann in the early 1970s.[39] This technique has found application in many areas similar to those discussed here, including the distribution of terpenes in crude oil fractions,[40] characterization of alkylbenzenes from coal extracts,[41] and identification of polyaromatic organosulfur compounds in fish and shellfish.[42] Its utility in arson analysis was noted early,[11,12] and recent papers in this area have dealt almost exclusively with this topic.[13-16]

In mass chromatography, complete mass spectra of the GC effluent are collected by computer every few seconds during the entire GC run. This ensures that no possibly important information about the sample is excluded and constitutes a critical difference from mass fragmentography (selected ion monitoring), in which only certain characteristic ions are monitored during the run.[43] Mass fragmentography is more applicable to the detection and quantitation of compounds in ultralow concentrations, which is not usually a serious concern in arson analysis.

Following the run, chromatograms for individual ions or groups of ions characteristic of specific compounds or classes of compounds are printed out in addition to the total ion chromatogram (TIC) of the sample. The TIC essentially mimics the output of a conventional GC, while the individual mass chromatograms are more like having highly selective detectors on a GC. Mass chromatography thus can identify the locations of compounds even at extremely low concentrations through many-fold enhancement of the signal-to-noise ratio (see Figure 2).

Mass chromatography alone, however, is not sufficient to identify the presence of single organic compounds, since the ions chosen could arise just as easily from extraneous compounds. Measuring the retention times of the chromatographic peaks obtained in each of the mass chromatograms is essential if one is to maximize the quality of information gained by this method. Biemann[39] has suggested the use of three internal standards of significantly different retention times to be co-injected with the sample for relative retention time measurements. Although this is desirable because of the length of most arson GC runs, it is often impractical due to the form of these samples (e.g., heated headspace). The choice of internal standards is important, since they should not duplicate compounds which could conceivably arise in the sample itself. Polychlorinated hydrocarbons offer as yet untested possibilities in this area, since they are commercially available, relatively inexpensive, and have fragmentation patterns characteristically different from those of most hydrocarbons. Alternatively, our laboratory has used a series of n-alkanes (C_5 to C_{16}) as an external standard run before each set of sample runs.[13] This option is useful if GC conditions are relatively stable from day to day.

The use of MS to characterize different compound types necessitates the choice of which ions to monitor. Gasoline, one of the most complex chemical mixtures of all the common accelerants, consists basically of only six different types of compounds: paraffins (saturated aliphatics), olefins, cycloparaffins, alkylbenzenes, alkenylbenzenes, and polynuclear aromatic hydrocarbons. For ease of identification, the polynuclear aromatics can be subdivided into the naphthalenes, indenes (isomeric with the alkenylbenzenes), biphenyls (and isomeric acenaphthenes), and anthracenes (phenanthrenes). Compounds of higher molecular weight generally are not observed under GC

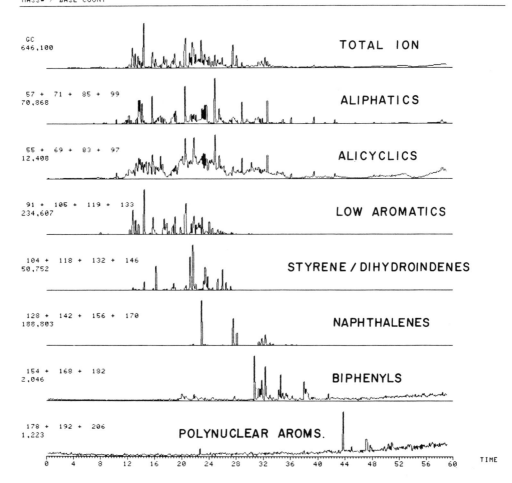

GC
646,100   TOTAL ION

57 + 71 + 85 + 99   ALIPHATICS
70,868

55 + 69 + 83 + 97   ALICYCLICS
12,408

91 + 105 + 119 + 133   LOW AROMATICS
234,607

104 + 118 + 132 + 146   STYRENE / DIHYDROINDENES
50,752

128 + 142 + 156 + 170   NAPHTHALENES
188,803

154 + 168 + 182   BIPHENYLS
2,046

178 + 192 + 206   POLYNUCLEAR AROMS.
1,223

0   4   8   12   16   20   24   28   32   36   40   44   48   52   56   60   TIME

FIGURE 2.   Family mass chromatograms from evaporated gasoline. (From Smith, R. M., *J. Forensic Sci.*,
28, 318, 1983. With permission.)

conditions normally used for routine arson analysis. Representative mass spectra of
compounds from several of these classes are shown in Figure 3.

Experience has shown that single ion mass chromatograms are often quite specific
for the characterization of polynuclear aromatic compounds,[11-14,16] since most of these
compounds have intense molecular ions and exhibit very little fragmentation even under EI conditions. This is not true for other compounds, however, and in these cases
multiple ion mass chromatograms (summing several different characteristic ion intensities) provide a more accurate picture.[44,45] We found that, especially for compounds
showing regular intense fragment ions (paraffins, alkylbenzenes, etc.), summing the
intensities of four different ions or ion ranges per mass chromatogram and plotting up
to ten such chromatograms on a single chart was particularly useful for most applications (Table 1).[13,14] Other authors have been successful using single ion chromatograms
exclusively.[16]

The list of ions in Table 1 also includes the terpenes, generally absent from petroleum products but abundant constituents of pine-derived solvents such as turpentine,
and ignores several types of compounds which could easily arise in arson accelerants.
Many oxygenated compounds, especially esters, alcohols, ethers, and some low-boiling

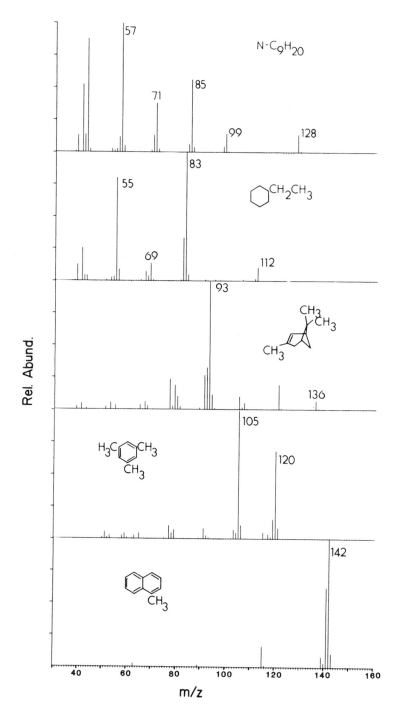

FIGURE 3.   Mass spectra of representative arson accelerant constituents.

ketones and aldehydes, do not give intense ions at these m/z values. As noted earlier, however, these compounds tend to occur in mixtures containing only a few intense components, where individual compound identification is relatively easy.

Kelly and Martz recently indicated several other ions for monitoring.[16] Most important among these was m/z 73 for the detection of methyl *t*-butyl ether, an octane

Table 1
MASS CHROMATOGRAPHIC FAMILIES

ALIPHATIC: C_4H_9, C_5H_{11}, C_6H_{13}, C_7H_{15}
57 71 85 99

ALICYCLIC/OLEFINIC: C_4H_7, C_5H_9, C_6H_{11}, C_7H_{13}
55 69 83 97

SIMPLE AROMATIC:

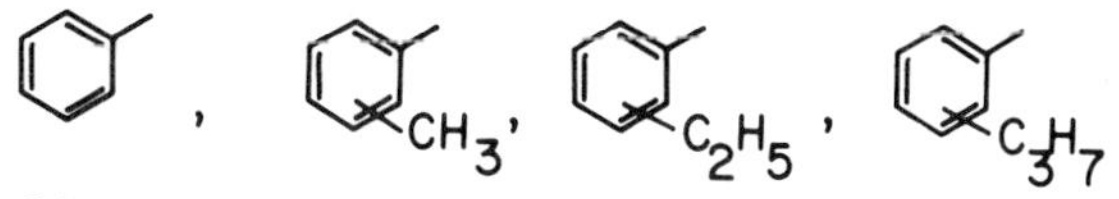

91 105 119 133

Also M.W. 92, 106, 120, 134, 148

STYRENE/DIHYDROINDENE:

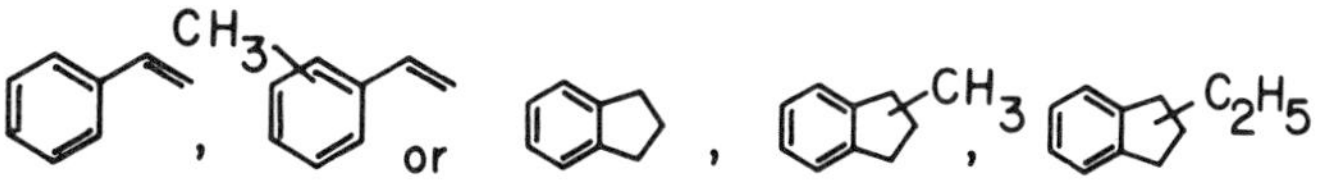

104 118 132 146

NAPHTHALENE:

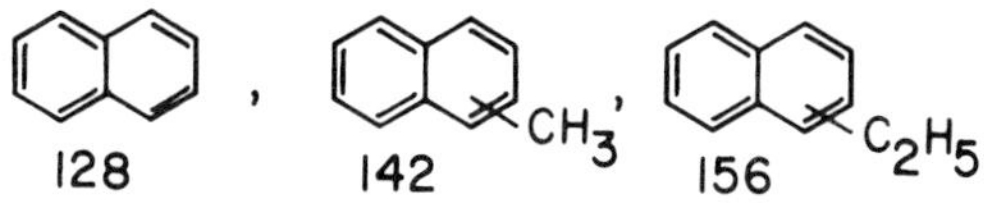

128 142 156

BIPHENYL:

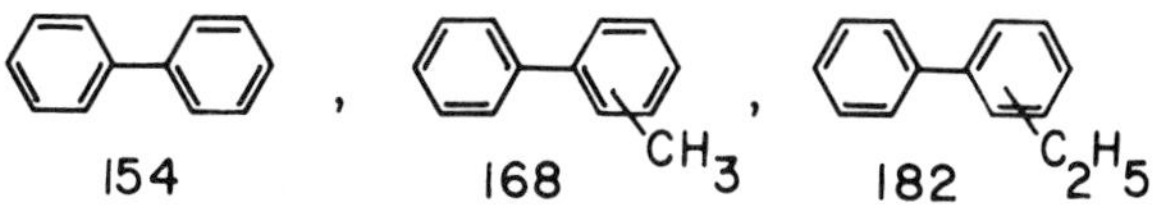

154 168 182

CONDENSED POLYNUCLEAR AROMATICS:

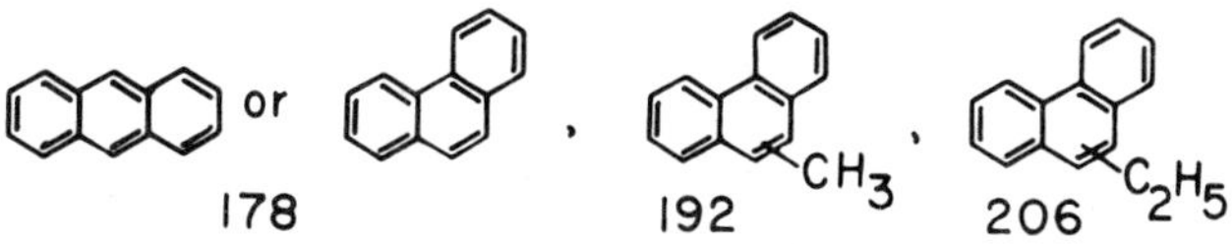

178 192 206

TERPENES: C_7H_9, $C_{10}H_{16}$ (e.g.

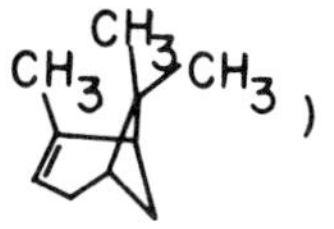

)

93 136

α- pinene

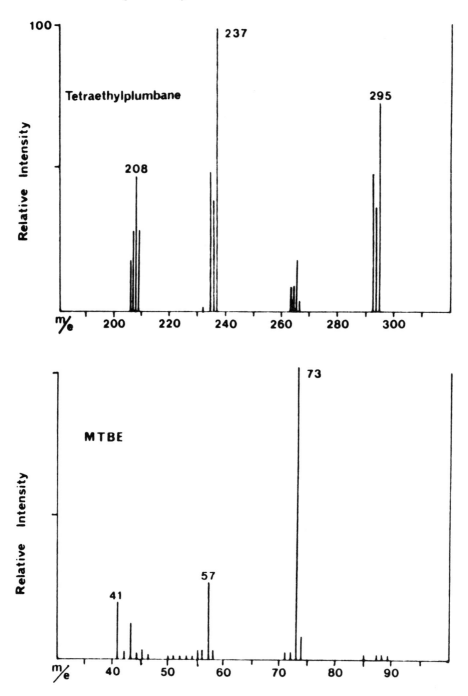

FIGURE 4.   Mass spectra of tetraethyllead and methyl *t*-butyl ether. (From Kelly, R. L. and Martz, R. M., *J. Forensic Sci.*, 29, 714, 1984. With permission.)

booster recently added to some unleaded gasolines, and the entire m/z 200 to 300 range for detection of tetraalkyl lead compounds in leaded gasoline (see Figure 4). Our experience has indicated that several extraneous compounds having molecular weights over 200 routinely occur in arson residues, so that narrower mass windows (such as m/z 235 to 237 and 293 to 295 for tetraethyl lead) may be necessary for consistent results.

In particular, butylated hydroxytoluene (BHT; 2,6-di-t-butyl-4-hydroxytoluene) is a common antioxidant with an intense fragment ion at m/z 205 which could interfere with the detection of free lead isotopes around m/z 206.

D. Form of the Data

The series of mass chromatograms generated by arson samples sometimes provides striking pictorial evidence for the presence of certain families of compounds.[13,14,16] The alkylnaphthalene and alkylbenzene series from gasoline are classic examples (Figure 5), while the n-alkanes from fuel oil are also noteworthy.[13,16] In each of these cases the location of individual compounds (or at least isomeric groups of compounds) was highlighted by plotting individual molecular ion chromatograms below the "family" mass chromatogram. For both the alkylbenzenes and the n-alkanes the molecular ion chromatograms were generated using different ions from those used for the "family" chromatograms, since the latter ions were chosen for sensitivity (greatest relative abundance) while the former were selected for compound specificity.[13,14]

Several families of compounds (the olefins and cycloalkanes in particular) consistently gave nondescript patterns by mass chromatography in nearly all samples, and chromatograms even for the usually cooperative groups sometimes yielded uninterpretable results. Thus it often was necessary, and generally desirable anyway, to identify as many of the important chromatographic peaks as possible by their mass spectra to ensure the presence of these compounds in the sample.

In addition to the visually identifiable chromatographic patterns formed by some families, the relative concentrations of various types of compounds gave characteristic information about the make-up of the sample.[13,14] This information, quantitatively crude though it may be, was available in our case on mass chromatographic printouts as the ion count of the largest peak in each mass chromatogram. Similar numbers are available from the printouts of other modern data systems. These numbers from individual samples were condensed into Table 2.

This table contains generally three kinds of information: (1) ion counts for each family of compounds, obtained from the "family" mass chromatograms and showing concentrations relative to each other and to the total ion count for the sample, (2) the presence or absence of a visual pattern of compounds typical for that family (a homologous series, as with the alkylnaphthalenes, or a characteristic collection of compounds, as with the terpenes), confirmed by comparison of GC retention times, library search, or both, and (3) the relative concentrations of compounds or isomeric groups of compounds as estimated either from the family mass chromatograms (relative peak heights times the "family" ion count) or from individual molecular ion chromatograms such as those in Figure 5.

III. SAMPLE CHARACTERIZATION

A. Standard Accelerants

Gasoline, because of its popularity as an accelerant and the relative ease with which it can be identified, has been examined repeatedly in the arson literature, often to the exclusion of other materials. As a result, a number of possible accelerants, particularly those derived from the light (naphthas, cigarette lighter fluids) and middle petroleum distillates (paint thinners and mineral spirits), have been somewhat neglected. Our own work in this area,[13] corroborated in the selected findings of others,[11,12,15,16] attempted to fill this gap through the examination of a variety of accelerants and their evaporated residues (see Table 2).

These results showed that, with few exceptions, mass chromatographic patterns pro-

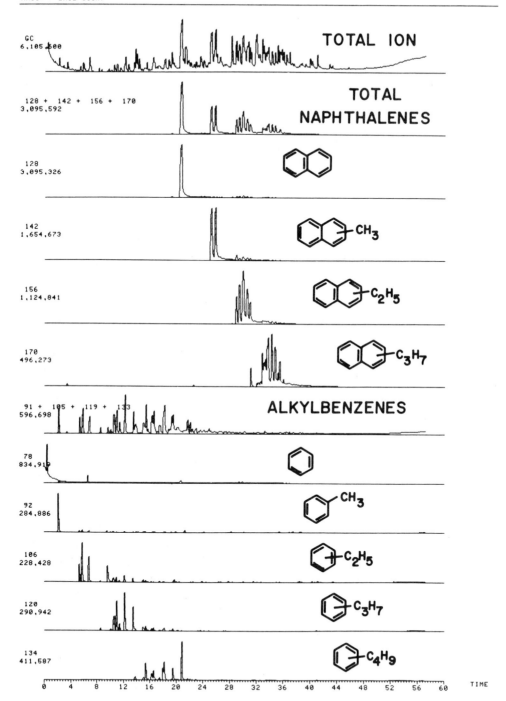

FIGURE 5.   Family and individual molecular ion chromatograms of alkylnaphthalenes and alkylbenzenes in highly evaporated gasoline. (From Smith, R. M., *Anal. Chem.*, 54, 1399A, 1982. With permission.)

duced by different accelerants were significantly and characteristically different. Arson residues containing measurable amounts of nearly unevaporated accelerants should be characterizable easily by this method. Results from evaporated accelerants were more difficult to interpret. Although even highly evaporated gasoline was identified easily by the assortment of middle- and high-boiling aromatics, several products had unusual compounds such as preservatives as major constituents of their residues.[13] It was not determined whether these compounds were truly distinctive. Much further work needs to be done to clarify this issue.

B. Background Materials

One of the greatest banes of the arson analyst is the peaks in a suspicious sample chromatogram which are intense enough to be significant, but whose pattern does not match that of any standard. The naive or incautious analyst might be tempted to characterize such samples in general, but nonetheless condemnatory, terms like "unidentified light petroleum distillate" or "possible mineral spirits". Juhala[46] correctly pointed out several years ago, however, that one must be extremely cautious about ascribing accelerant properties to these patterns without mass spectral characterization; even then, one must proceed with further caution.

Juhala's example illustrated a common occurrence in arson residue analysis — a fairly simple chromatogram containing several low-boiling peaks. Comparison with common solvent mixtures did not produce a match by GC alone. Identification of the peaks by GC/MS revealed the presence of benzene, toluene, ethylbenzene, and styrene, with styrene being the most intense peak. Although such a mixture undoubtedly is flammable, the question remained as to the origin of these compounds. Juhala postulated, based on several factors, that they must have arisen from the pyrolysis of styrene-containing polymers in the sample during the fire. Our laboratory has experienced a similar pattern of compounds (see Figure 6) on several occasions from a variety of sample types.[14,47] Indeed, the early identification reported by Zoro and Hadley[7] of an accelerant containing almost exclusively these same compounds must, in retrospect, be questioned in the absence of other information.

The problem of pattern contamination by background peaks arising via pyrolysis of structural or furnishing materials is a serious one — one which is particularly vexing without the use of GC/MS, since the analyst is often confronted with completely unknown peaks in a suspicious volatility range. Even an entire complex chromatogram can be due solely to background materials, as illustrated in Figure 7. This subject needs to be researched in much greater detail due to the enormous number of synthetic materials on the market today which are in common use. The complexity of the problem is inferred in Table 3, which shows the variety of compounds arising from pyrolysis of just a small number of materials which conceivably could be present in a typical arson residue.

C. Samples from the Real World

Identification of an arson accelerant from debris submitted as criminal evidence is almost entirely a matter of judgement, regardless of the sophistication of the methods used. The strongest reason for using GC/MS in this process is simply to give the analyst as much information as possible upon which to base this judgement. Each sample is truly unique and makes its own peculiar demands for handling, analysis, and interpretation. It is probably for this reason that the discussion of such samples has been noticeably rare in the forensic literature. Two examples are given below to illustrate the method and the problems involved.

Sample 1 was charred carpeting obtained from a fire involving a potential homicide.

Table 2

MASS CHROMATOGRAPHIC ANALYSIS OF STANDARD VOLATILES

Sample	Total ion count[a]	Aliphatics 57 + 71 + 85 + 99	n-Alkanes present	n-Alkanes maximum	Allicyclics 55 + 69 + 83 + 97	Low aromatics 91 + 105 + 119 + 133	Benzene 78	Toluene 92	Xylenes 106	C_3-Aromatics 120	C_4-Aromatics 134
Regular gasoline	2000	490 (24.5)[b]	5—16+	10—11	320 (16.0)	700 (35.0)	256 (12.8)	442 (22.1)	403 (20.2)	331 (16.6)	184 (9.2)
Fuel oil (#1)	331	148 (44.7)	8—17	11—13	25 (7.6)	23 (6.9)	—	1 (0.3)	5 (1.5)	23 (6.9)	14 (4.2)
Diesel fuel	6297	2272 (36.0)	9—20+	12—15	535 (8.5)	218 (3.5)	4 (.07)	28 (0.4)	53 (0.8)	137 (2.2)	87 (1.4)
Aviation fuel	6279	2011 (32.0)	5—9	5—6	N.A.[d]	1041 (16.6)	89 (1.4)	1040 (16.6)	600 (9.6)	400 (6.3)	200 (3.2)
Gulf-Lite™ Charcoal lighter	396	223 (56.3)	(8—16)[c]	—	19 (4.8)	N.P.	—	—	—	—	—
Coleman™ stove and lantern fuel	5457	1868 (34.3)	5—11	7—9	2048 (37.6)	336 (6.2)	80 (1.5)	176 (3.2)	189 (3.5)	41 (0.8)	7 (0.1)
Deodorine paint thinner	1805	697 (38.6)	8—12, 14, 16	10—11	320 (17.7)	253 (14.0)	—	—	9 (0.5)	147 (8.1)	47 (2.6)
Thinolene paint thinner	551	229 (41.6)	8—13	10—11	126 (22.8)	91 (16.5)	0.5 (0.05)	4 (0.7)	7 (1.3)	52 (9.4)	21 (3.8)
Benzine (low-boiling naphtha)	322	110 (34.2)	(7—9)[c]	—	129 (40.1)	1.4 (0.4)	—	0.5 (0.2)	1.4 (0.4)	—	—
Ronsonol™ lighter fluid (low-boiling naphtha)	235	81 (34.5)	7—9	8	64 (27.2)	15.2 (6.5)	—	15.2 (6.5)	14.5 (6.2)	—	—
Jon-E-Warmer fluid (low-boiling naphtha)	2516	816 (32.4)	7—9	8	218 (8.7)	110 (4.4)	—	76 (3.0)	63 (2.5)	—	—
Mr. Thinzit™ turpentine	2234	N.P.	—	—	N.P.	N.P.	—	—	—	—	—
Evap. regular gasoline	646	70	8—19	11—12	12	234	—	—	20	234	160

Evap. aviation fuel	981	(10.8) 500 (51.1)	9—10	9—10	(1.9) 33 (3.4)	(36.2) 123 (12.5)	—	(3.1)	(36.2) 56 (5.7)	(24.8) 52 (5.3)
Evap. Wizard™ charcoal lighter	249	97 (38.9)	12, 14, 16, 18	14, 16	20 (8.0)	N.P.	—	—	—	—
Evap. Coleman™ fuel/	6123	2784 (45.5)	10—12	10—11	1015 (16.6)	120 (2.0)	—	—	16 (0.3)	58 (0.9)
Evap. deodorine paint thinner	852	388 (45.5)	10—12	10—11	124 (14.5)	117 (13.7)	—	—	—	50 (5.9)
Evap. Benzine (low-boiling naphtha)	1149	393 (34.2)	(7—9),* 14—20+	16—18	305 (26.5)	N.P.	—	—	—	—
Evap. Ronsonol™ (low-boiling naphtha)	4732	367 (7.8)	7—20+	13—14	254 (5.4)	413 (8.7)	—	—	N.A.	N.A.
Evap. Jon-E Warmer fluid (low-boiling naphtha)/	1740	551 (31.7)	7—8, 11—20	16—17	163 (9.4)	63 (3.6)	0.5 (0.03)	1.1 (0.06)	N.A.	N.A.

Table 2 (continued)

MASS CHROMATOGRAPHIC ANALYSIS OF STANDARD VOLATILES

C_5-Aromatics 148	Terpenes 93 + 136	Naphthalene 128	Me-Naphthalenes 142	C_2-Naphthalenes 156	C_3-Naphthalenes 170	Styrene 104	Methylstyrenes 118	C_2-Styrenes 132	C_3-Styrenes 146	Anthracene 178	Me-Anthracenes 192	C_3-Anthracenes 206
77 (3.9)	N.P.ᶜ	311 (15.6)	311 (15.6)	100 (5.0)	5 (0.5)	25 (1.3)	171 (8.6)	130 (6.5)	70 (3.5)	—	—	—
8 (2.4)	N.P.	9 (2.7)	13 (3.9)	10 (3.0)	3 (0.9)	N.P.	N.P.	N.P.	N.P.	—	—	—
48 (0.8)	N.P.	131 (2.1)	498 (7.9)	480 (7.6)	170 (2.7)	N.P.	N.P.	40 (0.6)	153 (2.4)	—	—	—
—	N.P.	96 (1.5)	92 (1.2)	16 (0.3)	2 (0.04)	54 (0.9)	19 (0.3)	11 (0.2)	3 (0.07)	—	—	—
—	N.P.	—	N.P.	—	N.P.	N.P.	—	—	—	—	—	—
11 (0.6)	N.P.	N.P.	N.P.	N.P.	N.P.	N.P.	N.P.	N.P.	N.P.	—	—	—
6 (1.1)	N.P.	N.P.	N.P.	N.P.	N.P.	N.P.	N.P.	N.P.	N.P.	—	—	—
—	N.P.	11 (2.0)	3 (0.5)	—	—	N.P.	N.P.	N.P.	N.P.	—	—	—
—	N.P.	N.P.	N.P.	N.P.	N.P.	N.P.	N.P.	N.P.	N.P.	—	—	—
—	N.P.	N.P.	N.P.	N.P.	N.P.	N.P.	N.P.	N.P.	N.P.	—	—	—
—	N.P.	N.P.	N.P.	N.P.	N.P.	N.P.	N.P.	N.P.	N.P.	—	—	—
—	647 (29.0)	—	—	—	—	—	—	—	—	—	—	—
80 (12.4)	N.P.	188 (29.1)	110 (17.0)	39 (6.0)	4 (0.7)	—	28 (4.3)	39 (6.0)	5 (0.8)	1.2 (0.2)	0.4 (0.06)	0.2 (.03)
8 (0.8)	N.P.	143 (14.6)	130 (13.3)	50 (5.1)	10 (1.0)	—	5 (0.5)	17 (1.7)	6 (0.6)	7 (0.7)	4 (0.4)	1 (0.1)
—	N.A.	N.P.	N.P.	N.P.	N.P.	N.P.	N.P.	N.P.	N.P.	—	—	—
30 (0.5)	N.P.	N.P.	N.P.	N.P.	N.P.	N.P.*	N.P.	N.P.	N.P.	—	—	—

C_5-Aromatics 148	Terpenes 93 + 136	Naphthalene 128	Me-Naphthalenes 142	C_2-Naphthalenes 156	C_3-Naphthalenes 170	Styrene 104	Methylstyrenes 118	C_2-Styrenes 132	C_3-Styrenes 146	Anthracene 178	Me-Anthracenes 192	C_2-Anthracenes 206
117 (13.7)	N.P.	N.P.	N.P.	N.P.	N.P.	N.P.	N.P.	N.P.	N.P.	—	—	—
—	N.P.	N.P.	N.P.	N.P.	N.P.	N.P.	N.P.	N.P.	N.P.	50 (4.4)	6 (0.5)	1.5 (0.1)
N.A.	850[h] (18.0)	20 (0.4)	33 (0.7)	40 (0.8)	12 (0.3)	N.P.	N.P.	N.P.	N.P.	23 (0.5)	5.5 (0.1)	2 (0.04)
N.A.	N.P.	4 (0.2)	8 (0.4)	26 (1.5)	38 (2.2)	N.P.	N.P.	N.P.	N.P.	160 (9.2)	24 (1.4)	5 (0.3)

[a] All ion counts are given in thousands of ions.

[b] Numbers in parentheses show percentage of total ion count.

[c] N.P. — No pattern to indicate the presence of specific compounds.

[d] N.A. — Data not available for this compound or group.

[e] No n-alkanes observed; general range of branched alkanes shown.

[f] An important constituent of these samples was butylated hydroxytoluene (BHT), a common anti-oxidant (220 mol wt; ret. time about 36 min).

[g] A compound giving a large fragment ion at m/z 104 proved to be phenylcyclohexane.

[h] The major peak in the total ion chromatogram was an unidentified terpene alcohol; several aromatic esters, aldehydes, and sulfonamides were present also.

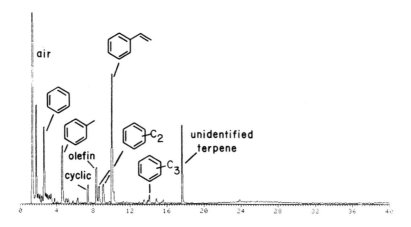

FIGURE 6.   Chromatogram from standard burned carpeting, showing distribution of styrene and low molecular weight alkylbenzenes. (From Smith, R. M., *Anal. Chem.*, 54, 1399A, 1982. With permission.)

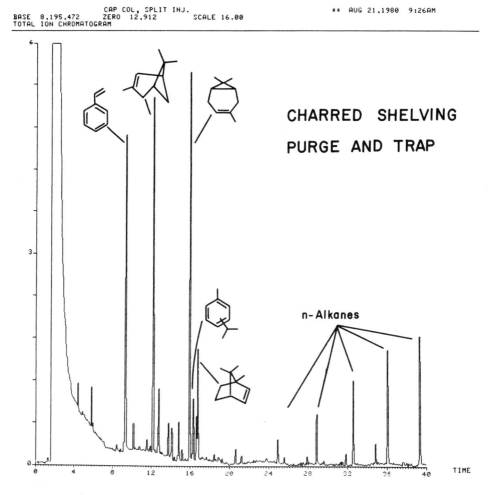

FIGURE 7.   Chromatogram from charred shelving in closet of suspect fire. Terpenoid compounds arise from wood in shelving material; high molecular weight *n*-alkanes may come from waxes. Intense peak at a retention time of 2 min is carbon disulfide from purge-and-trap procedure. (From Smith, R. M., *Anal. Chem.*, 54, 1399A, 1982. With permission.)

Table 3
PYROLYSIS PRODUCTS OF VARIOUS MATERIALS

Compound	Origin	Ref.
Styrene	Polystyrene	46, 53
	Acrylic paints	48, 52
	Adhesives	49
	Phenyl polymers	51
Benzene	Nitrile rubber adhesives	49
	PVC	50
	Phenyl polymers	51
	Polystyrene	46
Toluene	PVC	50
	Styrene-butadiene adhesives	49
	Polystyrene	46, 53
Xylene	PVC	50
Ethylbenzene	Polystyrene	46
C_3-Alkylbenzenes	Phenyl polymers	51
	Polystyrene	53
Naphthalene	PVC	50
	Phenyl ethers	51
Indenes	Styrene-butadiene adhesives	49
	Phenyl ethers	51
Dipentene	Natural rubber adhesives	49
	Phenyl polymers	51
Alkanes, alkenes, and cycloalkanes (C_4 to C_9)	Adhesives	49
	PVC	50
	Phenyl polymers	51
Acetone	Adhesives	49
	PVC	50
Acrylate and methacrylate esters	Acrylic paints and fibers	48, 52
	Acrylic adhesives	49
Low mol wt alcohols	Acrylic-based adhesives	49
	Some paints	52
Chlorinated alkanes	PVC	50

The analyst who first screened this sample by GC alone (Figure 8A) tentatively characterized it as an unidentified middle petroleum distillate, probably in the mineral spirits range. That an actual accelerant was present was strongly indicated by a detectable gasoline-like odor from the debris.

GC/MS analysis of this sample was run under far from optimal conditions (100 μl injection of heated headspace on a 5-ft by 1/8-in. I.D. packed 3% OV-101 column), but the only effect this had on the analysis was to make individual mass spectral examination slightly more difficult. The set of "family" chromatographs, obtained from a temperature-programmed run (50°C for 8 min, then to 250°C over 52 min), is shown in Figure 9. Further examination of the alkylbenzene and naphthalene families gave the individual molecular ion chromatograms shown in Figure 10.

Pertinent mass chromatographic data, obtained as described previously, is condensed into Table 4. In this case, the presence and length of the n-alkane series was estimated by comparison of GC retention times. The presence of benzene, toluene, and the xylenes in the complicated m/z 78, 92, and 106 molecular ion chromatograms was difficult to determine because of the intense fragment ions at neighboring masses from the 120-mol wt and higher alkylbenzenes. However, the lowest retention time peaks in each case (located by arrows in Figure 10) were identified by individual spectrum examination as the desired compounds. The presence of actual alkylstyrenes in this sam-

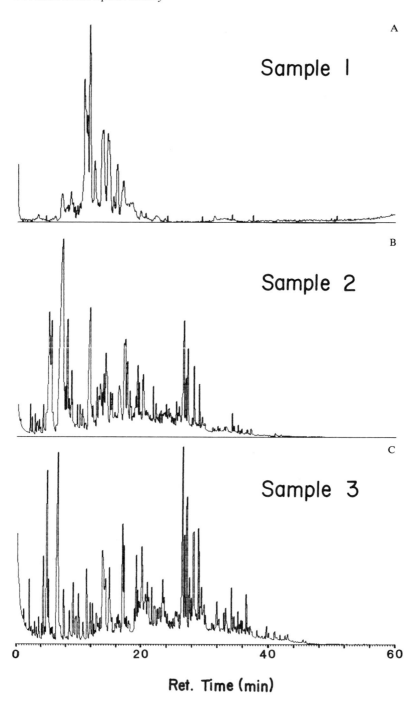

FIGURE 8. Chromatograms from three samples discussed in the text: (A) sample 1 — charred carpeting; (B) sample 2 — wood and debris, and (C) sample 3 — standard foam-backed carpeting. See also Table 4.

ple was confirmed by library search, but the peaks in the "terpene" mass chromatogram resulted from isotope peaks from the intense m/z 91 fragment ions of the 120-mol wt alkylbenzenes, the most intense peaks in the total ion chromatogram.

Data from the aliphatic, alicyclic, and alkylbenzene families for this sample fit rather classically those of a typical evaporated gasoline sample (compare with Table 2). Cor-

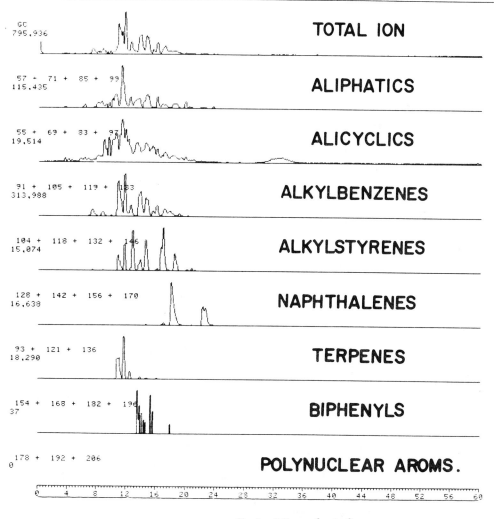

SCALE 1.00
MASS# / BASE COUNT   MULTIPLE MASS CHROMATOGRAM

FIGURE 9.   Family mass chromatograms from sample 1.

roborating this is the presence of the lower-boiling alkylnaphthalenes and middle al-kylstyrenes, albeit in relatively lower concentrations than those observed in the stand-ard. The most disconcerting feature about the data was the absence of the alkyl-anthracenes; these had been easily discernible in the standard, which otherwise exhib-ited a similar range of compound volatility to that of this sample. However, the fact that this sample was run as heated headspace vapor necessarily prejudices it against the higher-boiling compounds,[20] whereas the standard had been run as a wet needle injec-tion. The same rationalization serves to explain the smaller relative concentrations of the alkylnaphthalenes.

Sample 2 is much more challenging and illustrates the necessity of characterizing the entire sample, rather than relying only on the presence of certain key compounds for identification. This sample consisted of charred wood and other debris from a suspi-cious fire at a ski lodge. The total ion chromatogram (Figure 8B) was obtained from

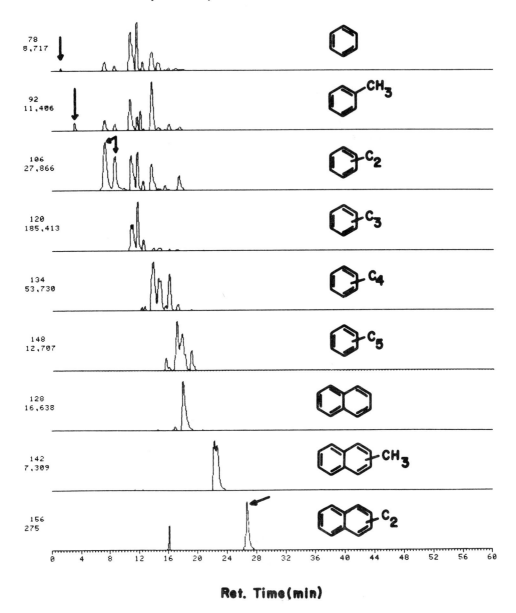

Ret. Time(min)

FIGURE 10.   Individual molecular ion chromatograms for alkylbenzenes and alkylnaphthalenes from sample 1.

CS$_2$ extract of the activated charcoal following "purge-and-trap" vapor collection. The GC conditions were similar to those for Sample 1, except that a 30-m SP 2250 WCOT capillary column was used instead of the packed column. In addition, the MS source filament was left off for the first 2 min of the run to allow the solvent to pass through the source undetected.

This chromatogram is interesting because of its complexity. It resembles that of gasoline in that it contains compounds having a wide range of volatilities, although the pattern itself does not match that of gasoline well enough to allow a positive characterization by GC alone. The "family" chromatograms (Figure 11) and other data (Table 4) for this sample give some surprising information. Although, on the one hand, this sample shows the relatively high concentrations of alkylbenzenes and alkylnapthalenes typical of gasoline, two pieces of information contradict this interpretation. First, the largest peak in the entire chromatogram is identified as styrene. In fact, the concentrations of the alkylstyrenes overall are much too large for those of gasoline. Second is the relatively low contribution due to aliphatic compounds, as well as the general lack of a typical *n*-alkane pattern. Indeed, comparison of the alicyclic and aliphatic mass chromatograms indicates that both may arise from alicyclic, rather than aliphatic, compounds. This is entirely atypical of gasoline.

The lack of alkylanthracenes in this sample is not particularly noteworthy, although the mode of vapor collection would have been expected to yield at least traces of these compounds from evaporated gasoline. The few terpenes observed are indeed real by individual spectrum identification, but are expected in small concentrations from samples containing wood.[13,14]

The question remains — what is this material? Is it an accelerant or a chance mixture of flammable hydrocarbons? The very large styrene peak in the total ion chromatogram gives the first indication that perhaps the material is *not* an accelerant. As we saw earlier, alkylstyrenes and alkylbenzenes are common pyrolysis products of many synthetic polymeric materials, and their presence in this sample at such high concentrations is more consistent with background materials. However, it came as a shocking surprise to us when an almost identical chromatogram (Figure 8C and Table 4) was obtained soon after this sample was run from standard, burned, foam-backed carpeting from an entirely unrelated case! Thus it appears that both of these chromatograms arise exclusively from the pyrolysis of "innocuous" materials, not from the presence of any accelerant.

D. Pyrolysis/Mass Spectrometry

A new approach to the characterization of accelerants by MS has been described recently by Voorhees and Tsao.[53,54] This technique, pyrolysis/MS uses Curie-point wires for vapor collection (in this work, from the effluents of combustion furnaces), followed by high-temperature desorption directly into the ion source of a mass spectrometer. Low-voltage (15 eV) ionization was used in order to minimize fragmentation, especially in experiments involving gasoline.

The resultant mass spectra, averaged over numerous repeated scans (Figure 12), are interesting superpositions of the mass spectra of all of the components of the sample. Most conspicuous in the gasoline spectrum are the various alkylnaphthalenes (m/z 128, 142, and 156). The large m/z 152 peak was not explained and has not been reported previously in arson residues, although it could conceivably result from high-temperature (450 to 500°C) formation of biphenylene from various aromatic compounds. Conspicuous by their near absence are the alkylbenzenes (m/z 106, 120, 134, etc.). This would seem to indicate extreme evaporation and weathering of the sample, although the simultaneous absence of alkylanthracenes (m/z 178, 192) contradicts this.

Sample characterization using this method depends not so much on individual compound identification as on the comparison of pyrolysis mass spectra from different samples by statistical evaluation. After complex factor analyses of the data, these authors concluded that the mass spectral patterns shown by pyrolyzed gasoline were repeatable and characterizable even in the presence of interfering substances such as pyrolyzed wool, wood, and synthetic polymers. The method has been used for the characterization of several other complex organic mixtures[55] and may be applicable to other accelerants as well.

The major advantages of pyrolysis/MS seem to lie in the ease and reliability of sample collection, small sample size (5 to 50 μg), and the rapidity and ease of data collection. On the other hand, the results obtained were not visually as convincing in

Table 4

MASS CHROMATOGRAPHIC DATA FROM SAMPLES

Sample	Total ion count[a]	Aliphatics 57 + 71 + 85 + 99	n-Alkanes present	n-Alkanes maximum	Alicyclics 55 + 69 + 83 + 97	Low aromatics 91 + 105 + 119 + 133	Benzene 78	Toluene 92	Xylenes 106	C3-Benzenes 120	C4-Benzenes 134
Charred carpeting	796	115 (14.5)[b]	8—14	9—10	19.5 (2.5)	314 (39.4)	+	2 (0.2)	28 (3.5)	185 (23.2)	54 (6.7)
Wood and debris	3446	409 (11.8)	(7—16)[c]	—	553 (16.0)	835 (24.2)	—	151 (4.4)	346 (10.0)	137 (4.0)	46 (1.3)
Foam-backed carpet	1624	326 (20.0)	(6—16)	—	382 (23.5)	171 (10.5)	—	114 (7.0)	64 (3.9)	58 (3.6)	23 (1.4)

C5-Benzenes 148	Terpenes 93 + 136	Naphthalene 128	Me-Naphthalenes 142	C2-Naphthalenes 156	C3-Naphthalenes 170	Styrene 104	Me-Styrenes 118	C2-Styrenes 132	C3-Styrenes 146	Anthracene 178	Me-Anthracenes 192	C2-Anthracenes 206
13 (1.6)	—	16 (2.0)	7.3 (0.9)	0.2 (+)	—	—	15 (1.9)	14 (1.8)	5 (0.7)	—	—	—
17 (0.5)	90 (2.6)	224 (7.1)	37 (1.1)	11 (0.3)	+	1331 (38.6)	548 (15.9)	33 (1.0)	35 (1.0)	—	—	—
10 (0.6)	57 (3.4)	164 (10.1)	18 (1.1)	2 (0.1)	+	559 (34.4)	139 (8.5)	19 (1.2)	0.8 (0.1)	—	—	—

[a]   All ion counts are given in thousands of ions.
[b]   Numbers in parentheses show percentage of total ion count.
[c]   No n-alkanes observed; general range of branched alkanes shown.

several of the examples given as one might like for legal purposes. Applications to actual arson residues are needed to establish the utility of this method.

IV. NEW DIRECTIONS

A. Tandem Mass Spectrometry

Perhaps the most promising recent addition to the repertoire of new analytical techniques in mass spectrometry is tandem mass spectrometry (MS/MS).[56,57] Basically, MS/MS by-passes the conventional forms of chromatography (GC, LC, etc.) by using mass selection as the chromatographic device. To achieve this, the tandem mass spectrometer consists of two mass analyzers connected sequentially, but separated by an intermediate area used as a collision chamber for fragmentation of the selected ions (see Figure 13). Although many combinations of mass analyzers (electrostatic, magnetic, or quadrupole) are possible, the "triple quadrupole" geometry has gained considerable popularity at the present time.[58]

In MS/MS the sample may be vaporized and ionized without significant prior workup, similar to the pyrolysis and thermal desorption technique just described.[55] For "site-specific" treatment of grossly heterogeneous samples, ionization can be induced by laser or similar methods.[59] The initial ionization, generally by methane chemical ionization (CI), gives a large preponderance of pseudomolecular ions, which are allowed to pass selectively through the first analyzer. These ions then are induced by collisions with an inert gas to fragment in the area between the analyzers, and the resulting fragment ions are separated in the second mass analyzer. Because of the option to scan either analyzer alone or both analyzers simultaneously, three basic types of MS/MS spectra are obtainable:

1. "Daughter" spectra — Analyzer 1 is maintained at a constant mass, while analyzer 2 is scanned to monitor the collisionally induced fragments from these ions. "Daughter" spectra are similar to conventional EI spectra and can be used for compound identification, especially if the ions passing analyzer 1 are characteristic pseudomolecular ions. Major drawbacks to these spectra at present include the need to generate "instrument-specific" spectral libraries (which may improve as the popularity of the technique increases), as well as the inability to separate isomeric compounds.

2. "Parent" spectra — Analyzer 2 is held at constant mass (looking, e.g., for a particular fragment ion), while analyzer 1 is scanned to determine which compounds (as their pseudomolecular ions) give rise to that particular fragment. "Parent" spectra thus are similar to single ion mass chromatograms, although with MS/MS the masses of both the fragment ion and its probable progenitors are determined in a single step.

3. Neutral-loss spectra — This is a more specialized application which allows searching for particular compound types, such as those whose molecular ions lose carbon monoxide or other specific small molecules. In this case, both analyzers are scanned simultaneously at the same rate, but with a constant mass offset reflecting the mass of the lost neutral fragment.

Although not applied specifically to the analysis of arson accelerants, MS/MS has been used for the identification of compounds in high-boiling coal liquids and other complex mixtures,[60] using both "parent" spectra to determine compound distribution and "daughter" spectra for compound identification. The characterization of arson residues, especially those involving less volatile compounds, may yield nicely to this technique.

B. FTMS

A second mass spectrometric technique with forensic possibilities is Fourier transform MS (FTMS).[61,62] This technique provides excellent sensitivity, very rapid scanning rates (<1 sec per scan), and concurrent high resolution ($>10,000$). FTMS grew out of ion cyclotron resonance MS and generates the mass spectrum by mathematical deconvolution of a nearly instantaneous picture of the simultaneous motions of all of the ions under fixed conditions.

Recent developments in this area have coupled capillary GC columns to an FTMS, thereby maximizing both mass spectral and chromatographic resolution.[61,63] Perhaps the *ne plus ultra* analytical instrument has been designed by Wilkins and co-workers in a system combining capillary GC with both FTIR and FTMS.[64] Identification of a lacquer thinner with this system has obvious ramifications in arson analysis.

V. SUMMARY AND CONCLUSIONS

Combined GC/MS has become an important tool for analyzing arson residues in

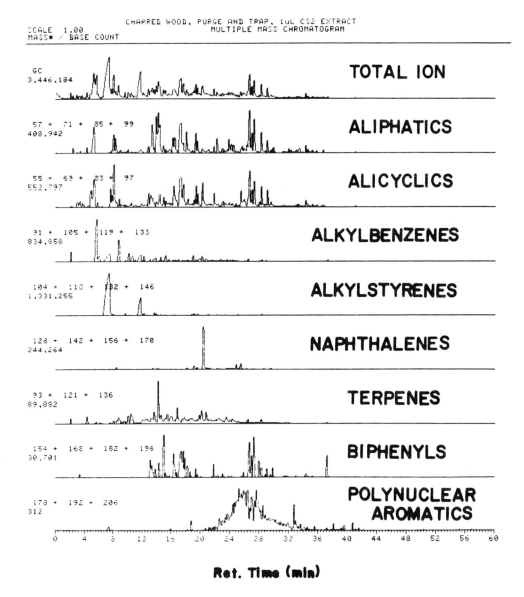

FIGURE 11.   Family mass chromatograms from sample 2.

recent years. Although plagued at first by a number of serious problems, new instrumentation offering high chromatographic resolution, faster scan rates, and sophisticated data manipulation has opened the door to several different approaches to the analysis of these materials. Characterization of samples by individual compound identification remains viable for simpler samples, while complex residues seem to yield best to mass chromatography. The emerging techniques of pyrolysis/MS, MS/MS, and FTMS potentially have exciting applications in this area.

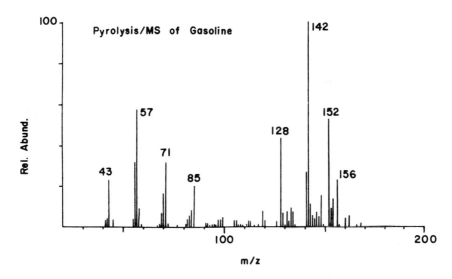

FIGURE 12. Pyrolysis mass spectrum of gasoline. (From Tsao, R. and Voorhees, K. J., *Anal. Chem.*, 56, 1339, 1984. With permission.)

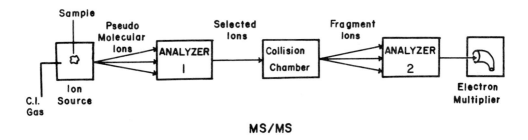

FIGURE 13. Schematic diagram of a MS/MS.

REFERENCES

1. Sieck, L. W., Fingerprinting and partial quantification of complex hydrocarbon mixtures by chemical ionization mass spectrometry, *Anal. Chem.*, 51, 128, 1979.
2. Camp, M. J., Analytical techniques in arson investigation, *Anal. Chem.*, 52, 422A, 1980.
3. Midkiff, C. R., Jr. and Washington, W. D., Gas chromatographic determination of traces of accelerants in physical evidence, *J. Assoc. Off. Anal. Chem.*, 55, 840, 1972.
4. Brettell, T. A. and Saferstein, R., Forensic science (review), *Anal. Chem.*, 55, 19R, 1983.
5. Choudhury, D. R. and Bush, B., Chromatographic-spectrometric identification of airborne polynuclear aromatic hydrocarbons, *Anal. Chem.*, 53, 1351, 1981.
6. Peaden, P. A., Lee, M. L., Hirata, Y., and Novotny, M., High-performance liquid chromatographic separation of high-molecular-weight polycyclic aromatic hydrocarbons in carbon black, *Anal. Chem.*, 52, 2268, 1980.
7. Zoro, J. A. and Hadley, K., Organic mass spectrometry in forensic science, *J. Forensic Sci. Soc.*, 16, 103, 1976.
8. Midkiff, C. R., Arson and explosive investigation, in *Forensic Science Handbook*, Saferstein, R., Ed., Prentice-Hall, Englewood Cliffs, N.J., 1982, chap. 6.

9. Smith, R. M., Some applications of GC/MS in the forensic laboratory, *Am. Lab.,* 10(5), 53, 1978.

10. Maucieri, L. A. and Bowden, J. P., Selection of residue and vapors from a simulated drug lab fire, *Arson Anal. Newslett.,* 4(2), 1, 1980.

11. Mach, M. A., Gas chromatography-mass spectrometry of a simulated arson residue using gasoline as an accelerant, *J. Forensic Sci.,* 22, 348, 1977.

12. Stone, I. C., Lomonte, J. N., Fletcher, L. A., and Lowry, W. T., Accelerant detection in fire residues, *J. Forensic Sci.,* 23, 78, 1978.

13. Smith, R. M., Mass chromatographic analysis of arson accelerants, *J. Forensic Sci.,* 28, 318, 1983.

14. Smith, R. M., Arson analysis by mass chromatography, *Anal. Chem.,* 54, 1399A, 1982.

15. Trimpe, M. A. and Tye, R., Mass fragmentography in arson analysis, *Arson Anal. Newslett.,* 7(2), 26, 1983.

16. Kelly, R. L. and Martz, R. M., Accelerant identification in fire debris by gas chromatography/mass spectrometry techniques, *J. Forensic Sci.,* 29, 714, 1984.

17. Higgins, K. M., Higgins, M. K., Oakes, C. L., and Braverman, S. F., High-speed extraction of accelerants from arson debris, *J. Forensic Sci.,* 29, 874, 1984.

18. Chrostowski, J. E. and Holmes, R. N., Collection and determination of accelerant vapors from arson debris, *Arson Anal. Newslett.,* 3(5), 1, 1979.

19. Saferstein, R. and Park, S. A., Application of dynamic headspace analysis to laboratory and field arson investigations, *J. Forensic Sci.,* 27, 484, 1982.

20. Nowicki, J. and Strock, C., Comparison of fire debris analysis techniques, *Arson Anal. Newslett.,* 7, 98, 1983.

21. Juhala, J. A., A method for adsorption of flammable vapors by direct insertion of activated charcoal into the debris sample, *Arson Anal. Newslett.,* 6(2), 32, 1982.

22. Andrasko, J., The collection and detection of accelerant vapors using porous polymers and Curie-point pyrolysis wires coated with active carbon, *J. Forensic Sci.,* 28, 330, 1983.

23. Armstrong, A. T. and Wittkower, R. S., Identification of accelerants in fire residues by capillary column gas chromatography, *J. Forensic Sci.,* 23, 662, 1978.

24. Willson, D., A unified scheme for the analysis of light petroleum products used as fire accelerants, *Forensic Sci.,* 10, 243, 1977.

25. Jensen, T. E., Kaminski, R., McVeety, B. D., Wozniak, T. J., and Hites, R. A., Coupling of fused silica capillary gas chromatographic columns to three mass spectrometers, *Anal. Chem.,* 54, 2388, 1982.

26. Settlage, J. and Jaeger, H., Advantages of fused silica capillary gas chromatography for GC/MS applications, *J. Chromatogr. Sci.,* 22, 192, 1984.

27. Dougherty, R. C., Negative chemical ionization mass spectrometry, *Anal. Chem.,* 53, 625A, 1981.

28. Guinther, C. A., Jr., Moss, R. D., and Thaman, R. N., The analysis and identification of weathered or fire-aged gasoline at various stages of evaporation, *Arson Anal. Newslett.,* 7, 1, 1983.

29. Merlé, F., Wiesler, D., Maskarinec, M. P., Novotny, M., Vassilaros, D. L., and Lee, M. L., Characterization of the basic fraction of marijuana smoke by capillary gas chromatography/mass spectrometry, *Anal. Chem.,* 53, 1929, 1981.

30. DeHaan, J. D., Report on the Congress of Criminalistics: Arson, *Arson Anal. Newslett.,* 3(1), 1, 1979.

31. Loscalzo, P. J., DeForest, P. R., and Chao, J. M., A study to determine the limit of detectability of gasoline vapor from simulated arson residues, *J. Forensic Sci.,* 25, 162, 1980.

32. Anon., Accelerant classification system, *Arson Anal. Newslett.,* 6(3), 57, 1982.

33. Sanders, W. N. and Maynard, J. B., Capillary gas chromatographic method for determining the C_3-C_{12} hydrocarbons in full-range motor gasolines, *Anal. Chem.,* 40, 527, 1968.

34. Gosselin, R. E., Hodge, H. C., Smith, R. P., and Gleason, M. N., *Clinical Toxicology of Commercial Products — Acute Poisoning,* Williams & Wilkins, Baltimore, 1976, sect. VI.

35. Gates, S. C., Smisko, M. J., Ashendel, C. L., Young, N. D., Holland, J. F., and Sweeney, C. C., Automated simultaneous qualitative and quantitative analysis of complex organic mixtures with a gas chromatography-mass spectrometry-computer system, *Anal. Chem.,* 50, 433, 1978.

36. Stenhagen, E., Abrahamsson, S., and McLafferty, F. W., *Registry of Mass Spectral Data,* Wiley-Interscience, New York, 1974.

37. Rosenthal, D., Theoretical limitations of gas chromatographic/mass spectrometric identification of multicomponent mixtures, *Anal. Chem.,* 54, 63, 1982.

38. Crawford, R. W., Hirschfeld, T., Sanborn, R. H., and Wong, C. M., Organic analysis with a combined capillary gas chromatograph/mass spectrometer/Fourier transform infrared spectrophotometer, *Anal. Chem.,* 54, 817, 1982.

39. Nau, H. and Biemann, K., Computer-assisted assignment of retention indices in gas chromatography-mass spectrometry and its application to mixtures of biological origin, *Anal. Chem.,* 46, 426, 1974.

40. Richardson, J. S. and Miller, D. E., Identification of dicyclic and tricyclic hydrocarbons in the saturate fraction of crude oil by gas chromatography/mass spectrometry, *Anal. Chem.*, 54, 765, 1982.
41. Gallegos, E. J., Alkylbenzenes derived from carotenes in coals by GC/MS, *J. Chromatogr. Sci.*, 19, 177, 1981.
42. Ogata, M. and Miyaki, Y., Gas chromatography combined with mass spectrometry for the identification of organic sulfur compounds in shellfish and fish, *J. Chromatogr. Sci.*, 18, 594, 1980.
43. Middleditch, B. S. and Desiderio, D. M., Comparison of selective ion monitoring and repetitive scanning during gas chromatography/mass spectrometry, *Anal. Chem.*, 45, 806, 1973.
44. Rawdon, M. G., Selective display of GC/MS data for compound class analysis and background correction, *J. Chromatogr. Sci.*, 22, 125, 1984.
45. Kuehl, D. W., Identification of trace contaminants in environmental samples by selected ion summation analysis of gas chromatographic-mass spectral data, *Anal. Chem.*, 49, 521, 1977.
46. Juhala, J. A., Determination of fire debris vapors using an acid stripping procedure with subsequent gas chromatographic and gas chromatography/mass spectrometry analysis, *Arson Anal. Newlett.*, 3(4), 1, 1979.
47. Haas, M. A., Olson, K. B., and Smith, R. M., unpublished data.
48. Saferstein, R. and Manura, J. J., Pyrolysis mass spectrometry — a new forensic science technique, *J. Forensic Sci.*, 22, 748, 1977.
49. Noble, W., Wheals, B. B., and Whitehouse, M. J., The characterization of adhesives by pyrolysis gas chromatography and infrared spectrophotometry, *Forensic Sci.*, 3, 163, 1974.
50. Gardner, R. O. and Browner, R. F., Determination of polymer pyrolysis products by gas chromatography and gas chromatography/mass spectrometry, *Anal. Chem.*, 52, 1360, 1980.
51. Jackson, M. T., Jr. and Walker, J. Q., Pyrolysis gas chromatography of phenyl polymers and phenyl ether, *Anal. Chem.*, 43, 74, 1971.
52. Wheals, B. B. and Noble, W., The pyrolysis gas chromatographic examination of car paint flakes as an aid to vehicle characterization, *J. Forensic Sci. Soc.*, 14, 23, 1974.
53. Tsao, R. and Voorhees, K. J., Analysis of smoke aerosols from nonflaming combustion by pyrolysis/mass spectrometry with pattern recognition, *Anal. Chem.*, 56, 368, 1984.
54. Tsao, R. and Voorhees, K. J., Fingerprinting of gasoline in combustion aerosols by pyrolysis-mass spectrometry with factor analysis, *Anal. Chem.*, 56, 1339, 1984.
55. Meuzelaar, H. L. C., Windig, W., Harper, A. M., Huff, S. M., McClennen, W. H., and Richards, J. M., Pyrolysis mass spectrometry of complex organic materials, *Science*, 226, 268, 1984.
56. McLafferty, F. W., Ed., *Tandem Mass Spectrometry*, John Wiley & Sons, New York, 1983.
57. McLafferty, F. W. and Bockhoff, F. M., Separation/identification system for complex mixtures using mass separation and mass spectral characterization, *Anal. Chem.*, 50, 69, 1978.
58. Yost, R. A. and Enke, C. G., Triple quadrupole mass spectrometry for direct mixture analysis and structure elucidation, *Anal. Chem.*, 51, 1251A, 1979.
59. Perchalski, R. J., Yost, R. A., and Wilder, B. J., Laser desorption chemical ionization mass spectrometry/mass spectrometry, *Anal. Chem.*, 55, 2002, 1983.
60. Cooks, R. G. and Glish, G. L., Mass spectrometry/mass spectrometry, *Chem. Eng. News*, 59(50), 40, 1981.
61. Wilkins, C. L. and Gross, M. L., Fourier transform mass spectrometry for analysis, *Anal. Chem.*, 53, 1661A, 1981.
62. Gross, M. L. and Rempel, D. L., Fourier transform mass spectrometry, *Science*, 226, 261, 1984.
63. Holland, J. F., Enke, C. G., Allison, J., Stults, J. T., Pinkston, J. D., Newcome, B., and Watson, J. T., Mass spectrometry on the chromatographic time scale: realistic expectations, *Anal. Chem.*, 55, 997A, 1983.
64. Wilkins, C. L., Giss, G. N., White, R. L., Brissey, G. M., and Onyiriuka, E. C., Mixture analysis by gas chromatography/Fourier transform infrared spectrometry/mass spectrometry, *Anal. Chem.*, 54, 2260, 1982.

Chapter 6

PYROLYSIS-MASS SPECTROMETRY OF SYNTHETIC POLYMERS

M. J. Whitehouse and B. B. Wheals

TABLE OF CONTENTS

I. INTRODUCTION

In principle, this chapter could encompass all the uses of mass spectrometry (MS) in examining synthetic polymers. However, it has been necessary to be selective, so, for example, the examination of extractable additives by gas chromatography/mass spectrometry (GC/MS) will not be dealt with. The main stress of current research has been on the use of MS for obtaining structural information or as a means of fingerprinting synthetic polymers. A major constraint on MS is the need for the sample to be in the gas phase or to yield gaseous ions. Obviously, polymers cannot usually be examined directly by MS without using some degradative technique to produce volatile products for examination. The most common method of achieving this is by pyrolysis under vacuum conditions or in the presence of an inert gas.

Many workers have utilized MS in the form of pyrolysis/gas chromatography/mass spectrometry (pyrolysis/GC/MS),[1-3] but in most cases once the products have been identified pyrolysis/GC is used for routine analysis. However, in the last decade, the direct combination of pyrolytic methods with mass spectrometry (pyrolysis/MS) has become a more popular alternative. The three techniques of pyrolysis/GC, pyrolysis/ MS, and pyrolysis/GC/MS are interrelated in the manner shown in Figure 1 and each has its own particular merits. Pyrolysis/GC benefits from only requiring inexpensive and widely available instrumentation, but suffers from the loss of information due to the nonelution of highly polar products and long analysis times. Pyrolysis/MS, on the other hand, has the advantage of speed of analysis and, in principle, all the products may be detected. Furthermore, it has the advantage of ease of use of automatic data-handling procedures due to a fixed mass scale. On the debit side, pyrolysis/MS requires complex and expensive instrumentation; interpretation of the data is not without difficulty due to the fact that a particular ion may arise from more than one product. With pyrolysis/GC/MS there is no such ambiguity in the results but large amounts of information are involved when comparison between samples is required. A method of overcoming this problem is to sum the spectra to yield just one mass spectrum.[4] Nevertheless, pyrolysis/GC/MS still suffers from all the disadvantages of both pyrolysis/ GC and pyrolysis/MS and is best used for structural work on unknown samples.

For the purposes of this chapter, only those procedures which routinely incorporate MS as the end detection method are considered. Such methods fall into three groups:

1. Those yielding a composite mass pyrogram built up by averaging all the spectral scans acquired during the significant part of the analysis (e.g., pyrolysis/MS, pyrolysis/GC/MS with scan averaging).
2. Those giving mass spectra of evolved compounds as a function of time (e.g., thermal ramping).
3. Those generating a profile of the variation in specific ion or total ion abundance as a function of time (e.g., thermal ramping).

It appears that the first application of pyrolysis/MS to the examination of synthetic polymers was reported by Zemany[5] in 1952. Submilligram quantities of material were pyrolyzed by a resistively heated filament at temperatures up to 1700°C, after which the volatile products were drawn from the pyrolysis vessel into the mass spectrometer. Using this very simple method it was possible to obtain mass pyrograms which were sufficient to characterize homopolymers, copolymers, and complex alkyd resins. Indeed, it was possible to differentiate eight resins of the latter type on the basis of the relative intensities of selected ions. However, despite this successful demonstration of the potential of pyrolysis/MS, the technique was neglected in favor of the cheaper and

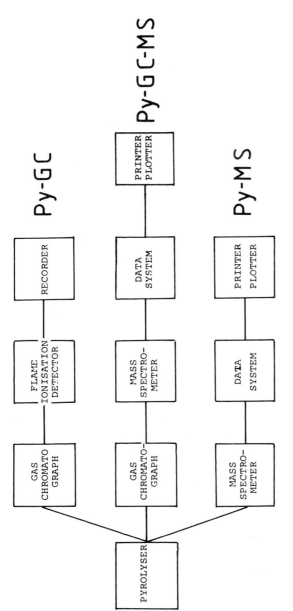

FIGURE 1. The interrelationship between the techniques of pyrolysis/GC, pyrolysis/GC/MS, and pyrolysis/MS.

simpler method of pyrolysis/GC. It was not until the early 1970s that the technique re-emerged largely as a result of the efforts of Meuzelaar and co-workers,[6] who established the conditions required to obtain rapid and reproducible mass pyrograms from complex materials. In the past 12 years, several groups of workers have developed various approaches to pyrolysis/MS which are applicable to synthetic polymers as well as other polymeric materials. The resurgence of the technique in this period has undoubtedly been due to the development of fast scanning spectrometers and online computers. Although these developments occurred to meet the requirements of GC/MS, they have been effectively employed in pyrolysis/MS for the rapid multiple scanning of the pyrolyzate and processing the large amount of data so produced. In a general sense, it is probably true to say that the technique of pyrolysis/MS is well established as is evidenced by the number of recent reviews[7-11] and books[12-14] devoted to the subject as well as biennial Gordon Conferences and international symposia.[15]

From the forensic point of view, however, pyrolysis techniques are still in their infancy, although Wheals[16,17] and Saferstein[18] have reviewed their development and the present state of the art. Few forensic laboratories were equipped with mass spectrometers a decade ago, and, if they were, the instruments were mainly used for the characterization of drugs and metabolites. Even today, only a small number of these laboratories utilize their mass spectrometers for polymer analysis, while infrared (IR) spectroscopy and pyrolysis/GC remain the methods of choice. This situation will not change unless it can be established that mass spectral techniques offer significant disadvantages (e.g., speed, greater reproducibility, higher discriminatory potential, lower unit cost, etc.). The case remains unproven, but the ease with which mass spectral data can be stored and manipulated in a computer, and the qualitative information that a mass spectrum contains, justify some investigation into the forensic potential of such techniques. The Metropolitan Police Forensic Science Laboratory in London has been studying pyrolysis/MS since 1975, and this review draws extensively on the experience we have gained since then.

In this chapter we deal with the instrumental aspects of pyrolysis/MS methods where the emphasis is placed on the pyrolysis procedures since it is assumed that the reader will be familiar with the MS requirements. This is followed by a discussion on aspects of sensitivity, reproducibility, and the treatment of results. In the final part we review applications of the technique specifically in the forensic field. Where possible, we have tried to keep the discussion relevant to the forensic scientist and, to this end, a few case examples have been included to illustrate its potential in this area.

II. INSTRUMENTAL ASPECTS OF PYROLYSIS MASS SPECTROMETRY

A. Pyrolysis Devices

A detailed discussion of pyrolysis devices will not be given here since an excellent review has been given by Irwin.[13] However, it should be noted that the majority of systems described involve pyrolysis *in vacuo* and therefore the temperature profiles observed under pyrolysis/GC conditions may not be the same as for pyrolysis/MS due to the considerably slower cooling processes.

1. Curie-Point Systems

Many pyrolysis/MS investigations have utilized Curie-point systems, but the first to be described was that of Meuzelaar and co-workers[6] at the FOM Institute in Amsterdam. A quartz reaction tube was connected to a stainless steel probe rod by a polytetrafluoroethylene (PTFE) holder which also retained the Curie-point wire onto which the sample was coated. When the probe was fully inserted, the sample area of the wire was at the center of the induction coil which was energized by a radio frequency (RF) generator (1.5 kW at 1.1 MHz). Pyrolysis was carried out with 510°C wires with a selectable duration of 0.2 to 4 sec. The pyrolyzate then entered the ion source of a quadrupole mass spectrometer through an intervening expansion chamber fitted with a pinhole leak. The expansion chamber was gold plated and maintained at 150°C to prevent condensation of the pyrolysis products. It served two purposes: firstly, to prevent contamination of the ion source by nonvolatile products, and, secondly, to broaden the pressure/time profile in the ion source in order to obtain an averaged mass spectrum. In fact, most of the condensable material is probably trapped in the unheated pyrolysis reaction tube. This arrangement represents a compromise between pyrolysis in the ion chamber (direct pyrolysis) which would exhibit high sensitivity but poor reproducibility due to contamination, and lower sensitivity and less information due to the loss of the higher molecular weight and structurally more informative products. A later modular version of this system[19] allowed easy removal of the expansion chamber and provided heating for the reaction tube. The potential contamination was reduced by heating the source (200°C) to slightly higher than the reaction tube (175°C). The pressure/time profile was extended to about 10 sec by the use of a slower heating rate of the wire (100°C/sec). This allowed efficient mass spectral scanning of the pyrolyzate and was sufficient to permit time-resolved profiles of the individual products to be obtained. It was claimed that this arrangement combines the advantages of flash pyrolysis with those of direct probe techniques, i.e., it minimizes char formation, secondary reactions, and loss of low-volatility products while retaining the valuable features of batch processing and easy automation. A commercial instrument (VG Pyromass 8-880),[20] which utilizes similar design concepts to that of Meuzelaar's but employing a small magnetic sector mass spectrometer has been described.

Schmid and Simon[21] have described a Curie-point pyrolysis/MS system utilizing a Knudsen reactor which combined the advantages of small (submicrogram) sample size and high vacuum pyrolysis. Ferromagnetic tubes (rather than wires) were held in a quartz tube attached to a direct insertion probe. The latter was inserted via a vacuum lock into an induction coil flanged directly onto one of the ports of the source housing. A heated quartz tube (5 mm I.D.) connected directly to the ion source of a double focusing mass spectrometer. The sample in solution or suspension was coated on the inside of the ferromagnetic tube by evaporation of the solvent. The temperature rise-time (TRT) was of the order of 200 msec for a 760°C tube and the maximum total ion current was observed after 0.5 sec. By the use of small samples the conditions are such that wall collisions predominate over intermolecular reactions. This was shown to be the case for the pyrolysis of 4-phenylbutyric acid since the formation of dibenzyl by recombination of phenyl radicals was virtually completely suppressed.

Hughes et al.[22] described simple modifications to a commercial GC/MS instrument to enable pyrolysis/MS to be obtained on polymeric materials of forensic interest. A Pye Curie-point pyrolyzer was mounted on the front of the gas chromatograph and purged with helium at a flow rate of 5 ml/min. Samples were pyrolyzed at 610°C for 15 sec and the pyrolyzate was swept into an empty column (45 cm × 2 mm I.D.) maintained at 200°C in the gas chromatograph. A make-up flow of helium (10 ml/min) was introduced at the end of the column and the mixture passed through a length of glass-lined stainless steel microbore tubing to the jet separator of a single-focusing mass spectrometer. The function of the empty column was to act as a buffer volume so that the pyrolyzate was extended sufficiently to allow 40 to 50 scans to be taken by a relatively slow scanning (3 sec/decade) magnetic instrument. Solid samples (down to 5 μg) were mounted by crimping in a loop formed by flattening the end of a Curie-

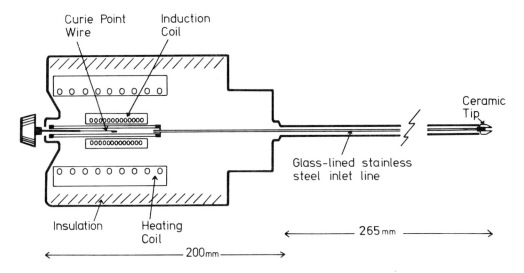

FIGURE 2.   Schematic of the VG Pyrolysis Probe.

point wire. A similar instrumental approach was utilized by Bracewell and Robertson[23] for the examination of soil extracts.

An alternative approach to using custom-built or modified commercial instrumentation is the use of a pyrolysis probe which can be used instead of a conventional direct-insertion probe. Such a unit (available from VG Analytical Ltd.), shown schematically in Figure 2, consists of a 1-mℓ glass sample volume surrounded by a 0.5-MHz induction coil connected to the ion source by 0.5-mm bore glass-lined tubing. The sample volume and inlet line may be operated at temperatures up to 230°C. The pyrolysis wire with the mounted sample is inserted in the center of the sample volume and, after pyrolysis, the products are drawn under vacuum through the inlet line into the mass spectrometer source. Under these conditions the pressure of the pyrolysis products decays at approximately 3% per second and permits 25 to 30 spectra to be taken at a rate of 2 sec per scan (35 to 450 amu) with a magnetic sector instrument. Although primarily designed for work in the electron impact (EI) mode, only simple modifications are required to allow the probe to be purged with a reagent gas to enable CI spectra to be obtained. Such a probe has been evaluated for forensic purposes[24,25] and its performance compared[26] with two other systems based on a Curie-point and Chemical Data Systems, Inc. (CDS) coil Pyroprobe,[27] respectively, and which were connected to a mass spectrometer in the manner reported by Hughes et al.[22]

2. Filament Systems

Developments in the design of the resistively heated filament pyrolyzers have led to TRTs comparable to those achievable with Curie-point systems. The most popular of these is the CDS Pyroprobe,[27] which comprises a booster-heated filament serving simultaneously as the heating element and as the temperature sensor. The filament is connected in a Wheatstone bridge-type circuit and the feedback control is such as to give reproducible temperature/time profiles. Using a platinum ribbon as the heating element onto which the sample is coated, heating rates of 0.1 to 20°C/msec from ambient to 1000°C can be achieved. Typical TRTs are 8 msec to 600°C and 17 msec to 1000°C. An alternative system is also achievable in the form of a coiled filament into which a quartz tube containing the sample is inserted. With this arrangement, the heating rates are slower (600 msec to 600°C) due to the higher thermal mass of the system. This disadvantage is offset by easier handling of insoluble and very small samples and

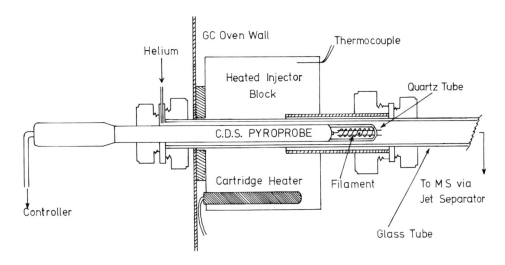

FIGURE 3. Schematic of the coupling of the CDS Pyroprobe to a mass spectrometer.

greater versatility. For example, samples may be repetitively pyrolyzed at progressively increasing temperatures or it may be fitted directly into a GC and even utilized as a direct-insertion probe for a mass spectrometer. At the Metropolitan Police Forensic Science Laboratory, a coiled filament-type Pyroprobe is used in conjunction with an empty glass tube in the GC oven of a GC/MS instrument. Figure 3 shows the detailed arrangement of the pyrolyzer in the injection port. The glass tube (23 cm × 10 mm I.D.) is connected by a reducing union to $1/_{16}$ glass-lined stainless steel tubing leading to the jet separator of a single-focusing mass spectrometer. The injection port, GC oven, and transfer line were all maintained at 230°C. After pyrolysis at 700°C for 20 sec the products were swept into the mass spectrometer by a flow of helium (6 ml/min). Mass spectra at 70 eV were taken every 2 sec over the range 25 to 250 amu and about 10 scans were averaged to yield the final mass pyrogram. The total analysis time was about 10 min, which is acceptable for forensic purposes. Hickman and Jane[26] have compared the performance of this system with that described by Hughes et al.[22] and the VG Pyrolysis Probe. The advantages of the CDS Pyroprobe in this arrangement were its superior reproducibility, higher sensitivity for the detection of the less volatile components, and the ease of handling of small samples. Its disadvantages were in the slowness of changing samples and the fact that the gas flows are interrupted during the sample introduction. Nevertheless, it has become the system of choice for the examination of casework samples at this laboratory.

The ribbon version of the CDS Pyroprobe was utilized with a modified quadrupole mass spectrometer for direct pyrolysis studies[28] By the use of scan rates up to 20,000 amu/sec, it was possible to obtain spectra from small samples (2 μg) for selected time intervals of 100 to 200 msec during the pyrolysis experiment. This instrument was primarily designed for kinetic studies, and preliminary observations showed that polystyrene decomposed under these conditions to yield essentially styrene monomer.

Saferstein and Manura[29] utilized a coil-type Pyroprobe to pyrolyze forensic samples (5 mg) into the heated (150°C) batch inlet system of a single-focusing magnetic sector instrument. After equilibration for 1 min the pyrolysis products were admitted into the ion source which was operated in the CI mode with isobutane as the reagent gas. Due to the nature of the sampling system it was necessary to use relatively large samples to obtain spectra of various acrylic polymers.

A rapidly heated (1000°C/sec) rhenium ribbon filament direct-insertion probe has been used for EI/CI studies of the pyrolysis of styrene polymers.[30] Israel et al.[31] have described the construction of a platinum coil pyrolyzer attached to a direct-insertion probe. This was controlled by a CDS Pyroprobe control unit and used for high-temperature pyrolysis/MS under CI conditions of technical polymers.

The technique of linear programmed thermal degradation mass spectrometry (LPTD/MS)[32] occupies, in an instrumental sense, an intermediate position between a filament pyrolyzer and the slow-heating probe systems to be described later. A U-shaped platinum filament attached to a direct-insertion probe acted as both sample heater and as a platinum resistance thermometer. The temperature of the filament was increased both linearly and stepwise from ambient to 600°C under computer control. The rate of temperature increase could be in the range 0.05 to 300°C/sec, but in the studies reported rates of 1 to 16°C/sec were used to enable time resolved mass spectra to be obtained.

3. Probe Systems

The direct-insertion probe offers the most convenient and simple way of obtaining pyrolysis/MS data without the need for specialized equipment. It can be used with mass spectrometers whose primary function may be for other than pyrolysis/MS work. In contrast to the previously described pulse-mode pyrolyzers, probes consist of an externally continuously heated tube into which the sample is introduced and belong to the class of furnace pyrolyzers. Due to the slow heating rates and the fact that the products must migrate through the residual sample before being released from the pyrolyzer, the production of secondary pyrolysis products is greatly increased. The technique is slow compared with flash pyrolysis — ca. 20 min compared with ca. 2 min — and as such not amenable to batch work or automation. Despite these disadvantages, this type of pyrolysis enables multidimensional data to be obtained and has been used for both characterization purposes and mechanistic studies.[33-40]

Most probe studies have involved temperatures up to 350°C with heating rates typically of 50°C/min but higher temperatures (650°C) can be achieved[40] using probes which have a water-cooled shaft to protect the vacuum seals. In all cases, the course of the pyrolysis can be followed by recording the total ion current (TIC) profile, and a change in the decomposition mode is often seen as multiple peaks in the profile. Furthermore, samples may be discriminated on the basis of these profiles, e.g., Figure 4 shows the TIC profiles of two different neoprenes. It will be appreciated that this technique is capable of generating large quantities of MS data compared with the single mass pyrograms obtained by flash pyrolysis methods. This can lead to an analogous problem to that in pyrolysis/GC/MS when one is required to compare the results obtained from a number of samples. The situation is made easier if specific ions can be identified as being characteristic of a particular product or series of products. This allows one to generate mass chromatograms in order to compare the relative amounts of the various products formed from one sample to the other.

Although the technique by itself would not necessarily meet the requirements for fast generation and comparison of data, it is a valuable adjunct to flash pyrolysis methods in that it can sometimes assist in the interpretation of the data from the latter. Furthermore, it may also provide additional data not observed with rapid pyrolysis methods. For example, this is the only way of readily detecting the pyrolysis products of nitrocellulose (NO_2 and NO) from nitrocellulose-modified alkyd resins.[41]

4. Thermogravimetry Systems

The use of thermogravimetry equipment linked directly to a mass spectrometer[42-45] yields data similar to those obtained by the use of a direct-insertion probe except that

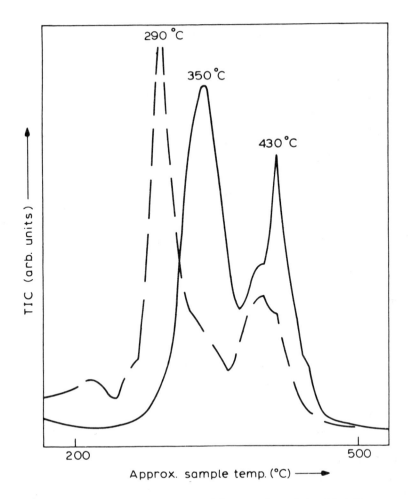

FIGURE 4. Total ion current traces of the ramped pyrolysis of (—) Neoprene GRT and (- - -) Neoprene WRT. (Magnification × 4.) (From Pidduck, A. J., *J. Anal. Appl. Pyrol.*, 7, 215, 1985. With permission.)

products may be related to a particular weight loss. In this technique the sample is slowly heated (5 to 50°C/min) on the pan of a thermobalance and the products are then usually swept into the mass spectrometer by a flow of helium via a heated transfer line and jet separator. In the method described by Smalldon et al.,[44] no intermediate separator was used and, although the flow of 10 ml/min of helium led to a high source pressure, the sensitivity and resolution was adequate for samples of the order of 500 μg. This sample requirement could be reduced to 10 μg if multiple ion detection methods were employed using ions which allow discrimination between samples. It is the high sample requirement and difficulties attendant upon coupling the thermogravimetric equipment to the mass spectrometer which detracts from the technique.

5. Laser Pyrolysis Systems

Pyrolysis of polymers by laser irradiation[46-50] is a particularly specialized technique, involving as it does the interaction of photons with the sample and processes which are only partly understood.[49] The mechanism of formation of polymer fragment ions by these methods depends on the wavelength of the laser used and the power density at the sample. Kistemaker et al.,[46] at the FOM Institute in Amsterdam, modified their previously described instrument to allow irradiation by a 20-W continuous CO_2 laser of a sample mounted at the aperture of the expansion chamber. Since the energy flux at the sample was only 0.5 W/mm², only neutral products entered the mass spectrometer where they were ionized by conventional electron impact (EI). Under these conditions they produced spectra similar to those obtained by Curie-point pyrolysis. Lum[47] has described a laser microprobe for the direct analysis of polymers which utilized relatively low-powered CW argon-ion and CO_2 lasers. Even though the energy at the sample could be increased to 10 W/mm² by the use of a focused beam, it was still necessary to ionize the neutrals using an electron beam. A commercial instrument (Leybold-Heraeus GmbH) has been developed for laser microprobe mass analysis (LAMMA) which employs a high-powered Nd:YAG Q-switched laser, the wavelength of which is frequency converted to yield UV photons.[48,49] At these power densities (10^7 W/mm²) the polymers yield low molecular weight ionized fragments on ionization. The original instrument (LAMMA 500) was primarily designed for use in the transmission mode with thin specimens, but a later development (LAMMA 1000) allows operation in the reflectance mode as well. A feature of these instruments is the ability to observe and manipulate the sample with a light microscope and then target the area of interest with a small "spotting" laser. A spatial resolution for analysis of 5 μm can be achieved which depends on the intensity of the irradiating laser. Characteristic spectra of both positive and negative ions from polymer samples can be obtained if the power levels are carefully controlled.[50] If too high a level is used then extensive pyrolysis and secondary reactions occur, leading to unspecific ion clusters. This type of instrumentation is unlikely to be used widely for analytical pyrolysis due to its high cost and highly specialized nature. Nevertheless, it will be invaluable for the examination of intractable materials (such as inclusions in polymers) or in laboratories which can also take advantage of its use for the detection of trace elements. However, the quantitative precision of the technique for this purpose has yet to be adequately demonstrated.

B. Sample Handling

As with any technique employed for the examination of small quantities of materials, in pyrolysis/MS it is important that careful attention be given to sample handling to avoid contamination and to obtain reproducible results. It goes almost without saying that all equipment with which the sample comes into contact should be scrupulously clean, i.e., glassware, Curie-point wires, quartz tubes, forceps, scalpels, etc. It has been shown, for example, that the mode of cleaning Curie-point wires[51] can affect the reproducibility of the results. Whatever method is chosen for this purpose it is important to keep rigorously to it in order to at least obtain good intralaboratory reproducibility.

The easiest samples to deal with are those which are readily soluble in a convenient and volatile solvent. In the case of Curie-point wires, these may be evenly coated by dipping the wire into the solution and then evaporating off the solvent. With filament pyrolyzers the solution should be transferred evenly in a thin film to the heating element using a microsyringe. Insoluble materials require more care and attention and are usually dealt with in one of two ways depending on the quantity of material available. When there is sufficient material it may be ground to a powder (under liquid nitrogen if necessary) and coated onto Curie-point wire or filament by forming a suspension in methanol or carbon disulfide. If the amount available is insufficient for this method (as is often the case with forensic samples) then other methods must be resorted to.

A small chip of paint, for example, can be examined directly with a Curie-point wire if it is crimped in a loop formed by first flattening the wire. This method has a number of disadvantages, the most important of which is the difficulty of manipulating small samples of 1 μg or less. Furthermore, if the loop is not satisfactorily crimped the sample may drop out, or heat transfer may be poor and ill-defined, which can affect reproducibility. The most convenient way of dealing with these types of samples, and in particular fibers, is to use the coil-type Pyroprobe where the sample is first loaded into a quartz tube. If possible, it should be cut into pieces of the appropriate size to permit replicate analyses, and, in the case of paint chips, additional layers should be removed with a scalpel. These procedures are made easier if a low-powered (ca. 40×) zoom microscope is used. After loading the sample in the center of the tube the latter should be inserted so that the specimen lies within the center of the coils.

Automation of sampling for pyrolysis/MS offers considerable advantages when large numbers of samples have to be dealt with. Such a system based on a Curie-point pyrolyzer and capable of dealing with 30 samples per hour has been described.[52] A more recent low-cost system[53] utilizes the Curie-point principle but, instead of wires, the samples are held in a V-section foil inserted in the mouth of a small rimless glass test tube. In practice this system is capable of a cycle time of 90 sec per sample and a maximum of 20 samples can be run in a batch. In principle the Pyroprobe-type pyrolyzers, either ribbon or coil type, are capable of being automated but no design has yet been reported. For forensic purposes there would appear to be little demand for automation since the sample load is insufficient to justify its consideration when deciding on suitable instrumentation.

C. Mass Spectrometry

It is not the intention to deal here with mass spectrometry (MS) in detail since it is assumed that the reader is reasonably familiar with the subject. Additional information may be obtained from References 13 and 54. However, it is necessary for the benefit of newcomers to the field to discuss some aspects of MS to aid them in the design of their experiments and the selection of appropriate instrumentation. The most important features which require amplification concern ionization, ion separation, ion detection, and computerization of the data. The latter is considered an aspect of MS since, today, almost invariably, mass spectrometers are sold as a package with a computer for both instrument control and data processing.

1. Ionization

By far the most common method of ionization involves the use of electron impact (EI) with the molecular beam of pyrolysis products. In conventional MS, an electron energy of 70 eV is used which is well in excess of the ionization potential of organic molecules (typically 10 to 15 eV). Dissipation of the excess energy among the vibrational modes of the newly formed ion leads to bond cleavages resulting in fragment ion formation. In pyrolysis/MS the superposition of this EI induced fragmentation for the multiplicity of products formed by pyrolysis results in a mass spectrum with some ion abundance at virtually every mass unit throughout the range scanned. While this can be of use for fingerprinting of synthetic polymers, it is undesirable for the elucidation of the structure of complex materials. Even without fragmentation there is no guarantee that a particular ion originates from a single pyrolysis product; in the situation of extensive fragmentation this probability increases dramatically. The fragmentation may be reduced by the use of ionization energies close to the ionization potential of the molecule. Generally speaking, "low" electron energies of the order of 10 to 15 eV have proved to be satisfactory,[55] although it should be borne in mind that some molecules may not be ionized at these energies. A further problem with the use of these low energies is the fact that small changes can lead to large changes in the spectra and can therefore affect reproducibility. For example, it has been shown[56] that changing the electron energy from 14.0 to 14.1 eV caused more than 10% difference in the average relative peak intensities. Furthermore, lowering the ionization energy can substantially reduce the ion currents obtained for a given amount of sample,[57] which can be of paramount importance for forensic work where the amount of material available is frequently limited. A major cause for concern is also the rather ill-defined nature of the electron energy on some instruments and the effects of source design. This was rather dramatically shown recently in an interlaboratory trial[58] involving six different instruments which yielded very erratic results for a standard compound at "low" electron energy. The choice of electron energy depends on the problem in hand; for structural purposes, "low" energies are to be preferred, but for fingerprinting the superior sensitivity and reproducibility of 70-eV ionization is recommended.

Simplification of pyrolysis/MS spectra may also be achieved by the use of chemical ionization (CI) methods with a variety of reagent gases. Although it has not yet become established as a routine method in pyrolysis/MS, a number of studies[24,25,29-31,59-63] have been made. The main reasons for its lack of widespread use are that, although the spectra may be more readily interpreted, the sensitivity and reproducibility are poorer than for EI. Furthermore, a large part of the lower end of the mass range is obscured by a high background due to the reagent gas. From the point of view of routine operation, prolonged use of CI can lead to rapid source deterioration and contamination of the pump oils. Despite these disadvantages, the technique can provide useful information for structural elucidation of pyrolysis products.[30,31,59-63] For some polymers, CI in the negative ion mode (NCI) can yield additional information, although only a few polymers produce spectra in both positive and negative ion modes.[61,62] The technique can also be used for the fingerprinting of samples of forensic interest. For example, Saferstein and Manura[29] have shown that acrylic-based automobile paints and acrylic fibers may be discriminated by the use of isobutane CI. It has been similarly shown[24] that isobutane CI can give superior discrimination of three acrylic paints than the use of low electron energy.

The related techniques of field ionization (FI) and field desorption (FD) have been employed to reduce molecular fragmentation of pyrolysis products. They differ in that with FI the pyrolysis is carried out by conventional means whereas with FD pyrolysis, desorption and ionization all occur at or near the emitter surface. In the case of FI there is still some residual fragmentation, but with FD the pyrolysis mass spectra are dominated by high molecular weight fragments which are mainly molecular ions and as such provide considerable information. Both methods have been employed[64-67] for studies on the pyrolysis products of synthetic polymers. However, due to the expensive and specialized nature of the equipment, they are unlikely to gain acceptance for fingerprinting purposes. Their main utility will remain the identification of pyrolysis products through accurate mass determination by high-resolution MS.

Fast atom bombardment (FAB), which has also been primarily used for obtaining mass spectra of high molecular weight biochemicals, has been shown[68,69] also to be capable of yielding MS data on synthetic polymers similar to that obtained by FD. Although the processes occurring at the specimen on the FAB target are little understood, the technique is simpler than either FI or FD and can be used with low resolving power instruments such as quadrupoles. It does, however, suffer from the disadvantage that there can be considerable interferences from the matrix material, e.g., glycerol, especially when dealing with small quantities.

It is claimed in a recent report[70] that photoionization MS in combination with pyrolysis/GC can overcome the deficiencies of pyrolysis/MS with low-voltage EI ionization. The energies of the photons are clearly defined, readily stabilized, and more efficient than EI ionization at lower energies. Furthermore, by use of different wavelength lamps, different classes may be selectively detected. As yet this method of ionization has not yet been applied to synthetic polymers.

2. Ion Separation

A number of methods are available for the mass analysis of the ionized pyrolysis fragments, and that which is used depends on a number of considerations. Among these are the purpose of the analysis, the mass range, scan speed, and the resolution required. Frequently, pyrolysis/MS studies have been carried out on instruments which have just happened to be available rather than being specifically chosen for the purpose in mind. It has been shown[58] that the qualitative nature of the mass pyrogram, at least for synthetic polymers, is more dependent on the pyrolysis process rather than on the subsequent MS.

Magnetic mass spectrometers have been the most widely used despite claims[71] of their unsuitability for pyrolysis/MS on the grounds of their lower scan speeds. Admittedly, most of the applications of this type of instrument have involved the use of slow direct-insertion probe pyrolysis where scan speeds are of less importance. Nevertheless, workable practical systems involving flash pyrolysis[20,22,23] have been used for routine fingerprinting work. Furthermore, there have been significant improvements in the scan speeds of these types of instruments and cycle times of 0.1 sec are now feasible.

Magnetic sector instruments have a number of advantages over quadrupole instruments. The source tuning is less susceptible to variations due to contamination and hence long-term reproducibility should be better. Other things being equal, it should be easier to obtain comparable results from instrument to instrument. Even single sector instruments can have a wider mass range, superior resolution, and exhibit no mass bias. A major advantage of magnetic instruments is that by the use of double-focusing techniques, high-resolution studies can give accurate mass information so that the elemental formula of ions can be determined.[72] However, in order to obtain accurate masses, it is necessary to maintain the pyrolyzate in the ion source for an extended period, which means that product distribution can be different from that obtained in flash pyrolysis. Furthermore, depending on the mass of the ion and the resolving power required, sensitivity may be considerably reduced and hence larger samples may be required than at low resolution. High-resolution instruments combined with FI and FD[64-67] have been used successfully for structural studies on pyrolysis products. However, accurate mass data alone is insufficient for the assignment of molecular structures and other methods must be resorted to. Mass analyzed-ion kinetic energy spectrometry (MIKES),[73] and exploiting metastable transitions by the use of linked scanning[74] are techniques which have been used for structural studies on the pyrolysis products and fragmentation routes of synthetic polymers.

Quadrupole mass spectrometers[75] have a number of advantages that make them a good choice for the basis of custom-built pyrolysis/MS instruments. They are capable of being very rapidly scanned (20,000 amu/sec), although this may be reduced in practice to ca. 2000 amu/sec by the amplifier bandwidth and the speed of data capture when used with a computer. Furthermore, since the ion source operates only at low voltages, contamination can more easily affect the source tuning and hence the final appearance of the spectrum. In addition, spectra are more likely to differ from instrument to instrument than is the case for magnetic mass spectrometers. It has been shown[76] that there were significant differences in the results obtained on the same sample with two instruments which had identical pyrolysis systems but different quadrupole assemblies. However, despite these limitations, pyrolysis/MS instruments based on quadrupole mass analyzers would seem to be perfectly satisfactory for fingerprinting work up to about mass 300 amu.

Time of flight mass spectrometers have found little use in routine pyrolysis/MS of synthetic polymers due to their complexity and limited mass range and resolving power. They may be used with advantage, however, for time-resolved studies using pulsed laser pyrolysis and in LAMMA instruments.

3. Ion Detection

Ion detection is most commonly achieved by the use of electron-multipliers in conjunction with analog DC amplifiers. A disadvantage of such systems is that the low bandwidths employed to suppress noise at high amplification can limit the scan speed, resolution, and sensitivity of the instrument. Obviously this would not be satisfactory for very small samples or when there is a need to record short-duration phenomena. Photographic recording with Mattauch-Herzog geometry mass spectrometers permits a wide dynamic range of registration and yields an integrated mass spectrum. However, this means of detection is less sensitive and convenient than electrical recording and tends to be used in specialist applications such as high-resolution studies with pyrolysis/FI or pyrolysis/FD. Ion counting is an alternative means of electrical recording which premits the very rapid scanning of mass spectra. It has been used with magnetic instruments[20,77] but it is in conjunction with quadrupoles[52] that it has been most beneficial in permitting the maximization of the analytical information, particularly for time-resolved studies.

4. Computerization

As in conventional MS, computers are in widespread use in pyrolysis/MS for instrumental control and data processing. It is not possible to go into detail and therefore the reader is referred to the excellent book on the subject by Chapman.[78] One of the roles of the computer is to control all the functions of the mass spectrometer and any auto-sampling equipment. While this role is very important and relieves the operator of much tedious work, its major function in pyrolysis/MS is to acquire repetitive scans of the pyrolyzate in order to obtain an averaged mass pyrogram. The computer also performs other important tasks in this respect, such as background subtraction and normalization of the spectra.

The ability of the computer to store large quantities of data can be exploited to build up libraries of mass pyrograms against which comparisons can be made automatically. Many of the commercial data systems and software available for use for pyrolysis/MS were originally designed with GC/MS in mind. They are usually based on small mini-computers with relatively small memories and the software for the comparison of spectra is somewhat limited to dealing only with fairly gross differences. With some effort it is possible to transfer py/MS data from such systems to main-frame computers so that small differences may be evaluated using some of the sophisticated statistical routines available in such common packages as, for example, SPSS (see Reference 79).

From the foregoing discussion of the instrumental requirements of pyrolysis/MS, it will be apparent that since no one system fulfills all needs, further studies are required to determine the optimum instrumental requirements for the various aspects of pyrolysis/MS.

III. ASSESSMENT OF PYROLYSIS/MS DATA

In this section we assess pyrolysis/MS data from the forensic scientist's point of view and consider sensitivity, reproducibility, and how the results can be treated.

A. Sensitivity

Many of the materials transferred during contact are minute, and in the case of particulate or fibrous samples the smaller they are, the greater their persistence on clothing.[80] Paint flakes submitted for analysis in our laboratory often have a mass of 5 μg or less, and in general it has been our conclusion that, provided the sample is large enough to be manipulated, analytical data can be generated from it. Samples are not routinely weighed, but for comparative analysis it is usual to make some effort to examine similar masses of suspect and control, and this can usually be ensured by pyrolyzing similar areas of the samples. An independent confirmation of this can be obtained if the total ion current generated from both are similar. In the pyrolysis/MS of alkyd paints, tests over the mass range of 1.5 to 20 μg have shown a high correlation between the mass of the sample and the TIC (correlation coefficient 0.95). Even when the sample masses are not identical we have found that with most paints spectral integrity is retained over the weight range previously mentioned. However, urethane-modified alkyds producing toluene di-isocyanate on pyrolysis (characterized by m/z 174) yielded anomalous results in that the relative ion abundance of m/z 174 increased as the sample size was reduced from 6 to 2 μg.

As might be expected, pyrolysis/MS with 70-eV EI ionization proved to be more sensitive than its CI counterpart, usually by a factor of 5 or more, although the sensitivity of the latter is very sample dependent. Pyrolysis/MS also shows a significantly higher sensitivity than those techniques relying on thermal ramping. Bearing in mind that most paints contain inorganic pigments in excess of 30% and that a 1-μg paint flake can yield a good-quality mass pyrogram, it is probably safe to say that there is no need to seek additional sensitivity.

B. Reproducibility

The use of any analytical results for forensic purposes carries with it a special responsibility, since their interpretation could ultimately influence an individual's liberty. Comparative analysis, which constitutes an important aspect of forensic contact trace examination, is impossible without some knowledge of the reproducibility and discriminatory potential of the methods used, and pyrolysis/MS, being a complex instrumental technique, requires careful control to be of any value. Windig and co-workers[51] investigated various factors which influence reproducibility in the analysis of biopolymers by pyrolysis/MS and concluded that careful attention to sample preparation, pyrolysis conditions, and mass spectral variables were essential for successful analysis. Hickman and Jane[26] compared paint samples by both pyrolysis/GC and pyrolysis/MS using filament and Curie-point pyrolyzers. They came to the following conclusions:

1. The major source of irreproducibility in pyrolysis/MS is the pyrolysis process.
2. Mass spectral tuning minimizes irreproducibility.
3. Some polymers pyrolyze more reproducibly than others.
4. Within a particular mass pyrogram some ions are produced more reproducibly than others.
5. Control of sample size is important.
6. Averaging multiple scans improves the reproducibility of many but not all ions.

Subsequent experience gained on the same system using the CDS Pyroprobe filament-type pyrolyzer has confirmed these conclusions.

There are three aspects to reproducibility which warrant attention viz. within-day and long-term reproducibility in a particular laboratory, and also the general problem of interlaboratory reproducibility. In order to quantify these various aspects it is convenient to concentrate on the variation of a few specific ions in the pyrograms of a control polymer, although others have used Euclidean distances based on multivariate analysis (see, e.g., Reference 51). The typical mass spectral data system presents ion intensity values normalized either to the base peak or the TIC. It should be appreciated that the relative standard deviations (RSD) derived from these are not identical. In our experience, normalization to TIC gives a lower RSD value, but for certain ions it is not

Table 1

MEAN INTENSITIES AND RELATIVE
STANDARD DEVIATIONS (% RSD)
FOR SELECTED IONS OF A
STANDARD ALKYD PAINT SAMPLE

m/z	77	91	104	105
Mean intensity on same day[a]	27	20	92	34
% RSD on same day	1.9	3.5	5.0	2.8
Mean intensity long term[b]	28	21	95	32
% RSD long term	7.1	13.1	7.0	8.7

[a]   Expressed as percent base peak (m/z 41).
[b]   Based on 1 year of data.

as good. Despite the slightly inferior data derived from intensities normalized to the base peak, the mass pyrograms which our customers have become accustomed to are expressed in this form. In the subsequent discussion the standard deviations quoted have been obtained from intensities normalized to the base peak.

In order to minimize long-term variation of pyrolysis/MS data, a control polymer is pyrolyzed at least once a week to ensure that certain ion abundances fall within a specified range. The sample used routinely is a white alkyd paint, since alkyds are the type of polymer system most frequently encountered in this laboratory. Tuning of the pyrolyzer and the mass spectrometer ensures that gross variations are minimized in the standard alkyd pyrogram, but even with this stringent control long-term variation is often three to four times worse than within-day variation. Typical RSDs for four ions in the standard alkyd are shown in Table 1.

The need to keep both types of variation to a minimum is crucial for forensic purposes. A match in suspect and control pyrograms (normally analyzed on the same day) can only be assessed by consulting a library of reference data. As was mentioned earlier in relation to sensitivity it is also a wise precaution to keep the sample size within a specified range to ensure the highest reproducibility.

Interlaboratory reproducibility is another feature which is desirable, but given the wide diversity of instrumentation and its high cost, little or no attempt has been made to standardize pyrolysis/MS procedures. A recent interlaboratory trial[58] brought out the variability in mass pyrograms of a variety of synthetic and natural polymers, and at the present time it is fair to say that only qualitative agreement is attainable. Most analysts using pyrolysis techniques accept that they have to build up in-house reference collections. Bearing in mind the speed of pyrolysis/MS analysis and the very small sample mass consumed, it is perhaps more realistic to aim to provide a common reference collection of samples rather than a common collection of pyrograms.

C. Treatment of Results

When a forensic scientist submits a polymeric sample for analysis, answers to the following four questions are required:

1.   What is the chemical nature of the sample?
2.   What is the likely use of such a sample?
3.   If two or more samples are submitted does the analytical result suggest that the samples are identical, similar, or different?

4.   If the suspect and the control samples come into the identical and similar categories how significant is this?

For the analyst attempting to use MS to answer such questions the task is twofold. First, there is the need to build up an extensive background collection of data about samples of known origin. Such a collection also needs frequent updating so that it reflects changing technological developments in the way polymers are used. Once this has been done the second task of interpreting results becomes possible provided an adequate knowledge of the reproducibility and discriminatory potential of the method has been acquired.

How then does the analyst set about treating his results? For the purposes of this chapter we will confine our attention to the way in which pyrolysis/MS data are handled — other methods generating similarly complex spectra can probably be treated in the same fashion. A diagram of the ways in which data can be treated is shown in Figure 5 from which it can be seen that they divide up into those utilizing the whole spectrum (mass pyrogram) or those in which only certain ion abundances are considered. Of the former methods, the visual comparison of pyrograms is the simplest approach and it provides a very effective way of screening data from large numbers of samples. The brain displays a remarkable capacity for this task and in this laboratory an experienced analyst is usually able to readily classify a polymer by this method. This is a valuable feature as it permits the background collection to be subdivided by polymer type, and it also means that the samples of different composition can be instantly recognized as such and a comparative analysis can be terminated at this stage.

An alternative to the visual comparison of mass pyrograms is to use a computer program to assess the degree of fit between them. This procedure yields a value known as the "fit factor" which is defined as follows:

$$\text{Fit Factor} = 1000 \times \left[1.0 - \frac{\Sigma(I_s - I_c)^2}{\Sigma(I_s^2 + I_c^2)} \right]$$

where I_s is the intensity of a given ion in pyrogram 1 and I_c is the intensity of a given ion in pyrogram 2.

Identical spectra yield a fit factor of 1000 while total dissimilarity is shown by a fit factor of 0. The fit factor approach gives equal weighting to each mass and will be most successful when large differences occur in occasional peaks rather than small differences in a large number of peaks. No correction for the differing reproducibilities of various ions is provided by the procedure, and this is a major weakness as this variation is known to occur. The main advantage of this method of comparing spectra is that it yields a single number which is convenient for reporting purposes. Experience has shown, however, that fit factors produced by a more selective use of the data are as good as those employing all results, and less computing time is involved.

Methods which are used more extensively for interpreting pyrolysis/MS data tend to be more selective than those outlined above. In practice, visual examination includes both the nonselective and selective approaches almost unconsciously. The primary visual screen serves to classify a polymer sample, and subsequent concentration on just a few ions permits further subclassification in many instances. Whether this can be achieved depends very much on the particular polymer system under investigation. In the case of alkyd paints, we have found that modifications due to toluene di-isocyanate, styrene, or vinyl toluene give rise to characteristic ions at m/z 174, 103, and 117, respectively. In these three instances the ions have a mass which is sufficiently removed from the region of greatest ion abundance (<m/z 100) to permit rapid visual recognition, and quite low levels of modification can be seen.

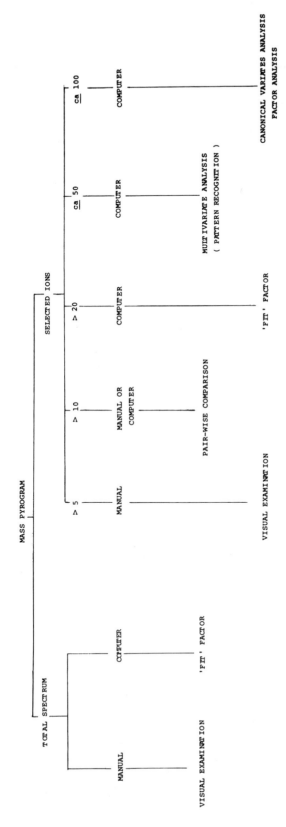

FIGURE 5.   Methods for the treatment of pyrolysis/MS results.

As pyrograms become more similar, and particularly when they lack any characteristic ions in the region above m/z 100, visual subclassification increases in difficulty. A particular weakness is that compensating for the natural variation in a pyrogram can be irksome. The irreproducibility of pyrolysis/MS does not lead to each ion varying to the same relative extent, and some ions may be three or four times less reproducible than others. The abundance of each ion in the mass pyrogram has to be considered as a point within a range, and the boundaries of this range varying from ion to ion. This means that mass pyrograms which may look slightly different may be statistically identical. When visual examination reaches this stage it is necessary to resort to some mathematical treatment to resolve the question of identity or nonidentity.

The simplest of the mathematical procedures is to undertake a pair-wise comparison of the pyrograms. A number of ions (usually less than ten) are selected on the basis of having a low level of correlation with each other and displaying a wide variation in abundance in the normal population of that particular polymer/material type — both these criteria are ascertained from reference data. To make a judgement as to whether or not two samples are similar, the measured ion abundance of each ion is taken to be the mean value of a range specific to that ion in that type of sample. Range overlap is considered to be a measure of similarity. The range is determined by the standard deviation of that ion in a control polymer of the same type. It is important to distinguish between same-day and longer-term variation, and to apply the appropriate value when conducting the pair-wise comparison. We normally set the range limits as $\pm 2\sigma$ and score overlap between two samples as plus and nonoverlap as minus. A comparison yielding all plus scores would be taken to indicate a similarity between the samples, but does not establish identity, for a nondiscriminatory analytical method or a very high incidence of chemically similar samples in the normal population would lead to a similar result. To make a judgment on the significance of a high plus score, it is necessary to refer back to the reference collection, and more specifically to look at data on samples of the same polymer class and application. Thus, in the context of white paints found to be alkyds by analysis (and known to be for decorative purposes rather than vehicle paints from the scene of crime), a pair-wise comparison with the data from all white decorative alkyd paints in the reference collection would be made. The standard deviation used to set the range limits in this instance would be the value for each ion found over the long term. Depending on the size of the reference collection and the efforts taken to ensure that it is representative of the "real" population of a specific type of sample, a judgment on the significance of the matching of suspect and control becomes possible. Samples which are indistinguishable from a high proportion of those in the reference collection can be considered to have a lower probability of being identical than those which are unique in character. In reporting analyses of that type we would state — "Samples A and B are based on . . . (polymer type) and their mass pyrograms were indistinguishable visually and by pair-wise examination of X key ions, using $\pm = 2\sigma$. Pair-wise comparison against the reference collection indicated that they were indistinguishable from Y% of the total number of samples N, having similar color, composition, and usage." The forensic scientist reporting the case has to draw his/her own conclusions about the significance of the matching data based on the values of Y and N, and his confidence in the method. This semistatistical approach to evaluating contact trace data is finding increased usage in this laboratory where every effort is being taken to reduce the subjectivity of all examinations.

The fit factor approach described earlier can also be used in a selective mode with various key ions being compared. The disadvantages previously mentioned still apply, but it can be conveniently used with more variables than the pair-wise comparison procedure but so far no comparative trial of the alternative procedures has been made.

Progressing still further into the realms of multivariate analysis, one begins to reach a point where mental assimilation of so many individual results becomes an impossible task. A bewildering array of statistical techniques for handling complex data are available and the only way to assess their value is to apply them to the task and see what happens. The number of variables that can be processed by a particular statistical package and the related demands of computer memory and time limit the option. In our laboratory, a "cluster analysis" procedure that will only accept 50 values per sample has been studied.[81] To reduce our mass pyrograms which cover the mass range 250 to 25 amu, it is necessary to reject over 150 values per pyrogram. Four ions associated with the air background (i.e., m/z 28, 32, 40, and 44) are the first to be discarded, and the remaining 221 ions are reduced by comparing data from 30 replicate analyses of a standard sample against that from 30 samples of the same type of material in the reference collection. The following steps are then taken in sequence:

1. Ions with intensities <5% of the base peak in both sets of data are rejected.
2. Ions in which the range of abundance is greater in the 30 replicates than in the 30 samples are rejected.
3. Ions in which the range of the abundance is similar in both the data sets are rejected.
4. Correlation coefficients for every pair of remaining ions in both sets of data are calculated, and in those pairs which are highly correlated (correlation coefficient >0.07) a single ion (usually that of lower abundance) is rejected.

This process usually permits over three quarters of the variables to be rejected and still leaves a representative selection for data processing. In the case of alkyl paints, clustering produces about eight subgroups, but the chemical significance of these groups has still not been deduced, nor has their long-term validity been established. At present, in this laboratory, the way in which pyrolysis/MS data are treated is still in a state of flux, but we hope ultimately to make a systematic comparison of the procedures outlined above. Where the computing facilities are available, procedures can be used to process pyrolysis/MS data with over 100 variables. For those interested in this type of treatment, recent publications[82-84] describe alternative methods.

IV. THE APPLICATION OF PYROLYSIS/MS IN FORENSIC SCIENCE

A. Polymer Identification and its Significance in Forensic Science

That every contact leaves a trace is an old and universally applicable maxim, but the materials transferred during this process fall into two categories. One group of contact traces consists of natural products such as body fluids, hair, plant materials, etc. whereas a second consists of manufactured materials. During the 20th century the development and exploitation of synthetic organic polymers has produced a remarkable array of products falling into the latter category. This has created many more problems as well as opportunities for the forensic scientist.

The type of polymeric samples submitted for pyrolysis/MS analysis in this laboratory in 1984 are shown in Table 2, which is typical of the pattern of submission over many years. Virtually all the samples examined arose from cases involving physical transfer of material (not always in trace amount), or from equipment thought to have been used or constructed for criminal purposes. Paint flakes, a common transfer material in break-ins and automobile crimes, are predictably of major importance. Adhesives can be present in bombs, and other devices used in the commission of crime, as can also plastic-covered wires and tapes. The low incidence of fibers among the

Table 2
SAMPLE TYPES SUBMITTED
FOR PYROLYSIS/MS ANALYSIS
AT THE METROPOLITAN
POLICE FORENSIC SCIENCE
LABORATORY IN 1984

Paints	59%
Adhesives	8%
Rubbers and foams	4%
Fibers	0.5%
Structural plastics	10%
Wire coverings and adhesive tapes	7%
Miscellaneous	11.5%

samples does not reflect the relative importance of this type of material, but is more attributable to the way in which this laboratory is structured and the barriers created between the divisions. In common with many other forensic laboratories, fiber examinations are conducted by forensic biologists, and they have evolved their own techniques which rely heavily on IR spectroscopy, microscopy, and chromatography of the extracted dyes. Whether pyrolysis/MS could be used more profitably for fiber classification has yet to be decided. Similar barriers exist in samples derived from "documents" casework; these are classified in the miscellaneous category in the table, but some excellent results have been achieved with photocopying toners, type correcting fluids, and adhesives present on documents. Submissions of items of this type are not high, and again this may be attributable to interdisciplinary attitudes which can distort the submission patterns.

Despite a substantial overall case load in this laboratory, the number of samples ultimately being submitted for pyrolysis/MS analysis is quite small and over the last 3 years has averaged about 370 per annum. It must be appreciated that the samples submitted for comparative instrumental analysis have usually been screened by other methods and only samples which cannot be discriminated by these procedures go forward for further analysis. Only rarely does polymer analysis produce evidence permitting a scientist to state that the probability of a suspect and control having come from the same scene is 1.0 although it can establish nonidentity with considerable certainty. In dealing with mass-produced materials such as commercial polymers, it has to be accepted that analytical identity of a suspect and control cannot prove they had a common origin. Much additional information is required to establish a firm link between such samples, not least being an extensive background knowledge of polymers and their incidence in particular circumstances. It is the acquisition of the latter by permitting a very extensive data collection to be established with speed that a technique such as pyrolysis/MS could prove most valuable.

B. The Use of Pyrolysis/MS in Forensic Science

The use of the technique in the forensic science laboratory differs significantly from that of other laboratories in respect to the nature of the sample and the information required from the analysis of it. The forensic sample is almost invariably limited in quantity, often insoluble and filled with inorganic materials, and maybe contaminated by other materials, such as dirt and grease. Besides being such intractable materials to deal with, there is often little information available as to its possible nature. The samples may arise in two distinct situations which require different types of answers to be provided. By far the most common is the comparison between samples such as paint,

Table 3
PAINT TYPES SUBMITTED
FOR PYROLYSIS/MS
ANALYSIS AT THE
METROPOLITAN POLICE
FORENSIC SCIENCE
LABORATORY IN 1984

Vehicle paints	47%
Household gloss paints	36%
Tool paints	12%
Emulsion paints	5%

fibers, adhesives, and plastics found at the scene, and with those samples on, or in the possession of, a suspect, in an attempt to link the two. In this situation it is necessary to show that the samples are identical and that there is a high probability that this could not have arisen by chance. This requires that the technique must have a high discriminating power and be capable of amassing considerable amounts of background data. The extent to which pyrolysis/MS can fulfill these requirements is illustrated later. A less common, but just as important, type of sample arises when there is nothing to compare it with, i.e., when no suspect has yet been apprehended. This might arise, for example, in a hit-and-run incident (when a flake of paint is left at the scene) or materials used in the manufacture of terrorist bombs. In these situations it is necessary, in order to help the police with their investigations, to provide as much information as possible, e.g., the year and make of the vehicle or a possible brand name for an adhesive.

C. Applications of Pyrolysis/MS
1. Paint

Paint constitutes one of the most important evidential materials dealt with by the forensic science laboratory and so it is treated here in some detail. As can be seen from Table 2, paint in this laboratory is submitted for pyrolysis/MS analysis more frequently than any other type of sample.

The breakdown by type (Table 3) shows that vehicle and decorative paints predominate, while tool and emulsion paints only arise occasionally. With the exception of emulsion paints, the majority are based on alkyd resins which are complex polyesters formed by copolymerizing phthalic anhydride, polyols, and saturated or unsaturated fatty acids. Dried films can be formed by further cross-linking taking place at double-bond sites in the fatty acid chain, as is the case with most decorative gloss paints. With vehicle paints, cross-linking may be promoted by the incorporation of a nitrogen-containing copolymer such as butylated melamine followed by stoving. Film formation in the case of air-drying spray paints such as "touch-up" paints is by simple solvent evaporation.

The mass pyrograms of alkyds, typical examples of which are shown in Figure 6A and B, are dominated by ions due to phthalic anhydride (m/z 104, 76, 50, and 148) although other ions present also reflect the "oil-length". The latter is the percentage of oil incorporated in the resin and is used to arbitrarily classify them as:

Short oil, <45%
Medium oil, ~45 to 65%
Long oil, >65%

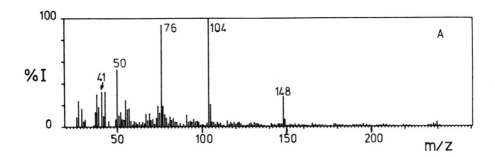

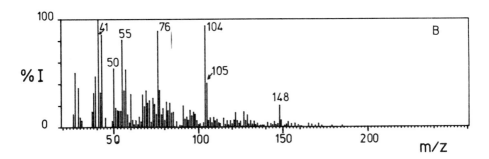

FIGURE 6. Mass pyrograms of a short oil (A) and a long oil (B) alkyd.

The short oil alkyds are predominantly used in paints for metal surfaces (e.g., vehicles and tools) which are stoved, while medium/long oil ones are utilized for air-drying decorative gloss finishes, mainly for wood surfaces. It has been shown in this laboratory that there is a good correlation (Figure 7) between the ratio of the intensities of ions characteristic of the oil (m/z 41 and 43) to those of the acid constituents (m/z 104 and 105). This information can be used to indicate the likely use of the paint in the absence of any chemical modification.

Alkyds may be modified with a wide variety of materials and some of the resulting mass pyrograms are shown in Figures 8 and 9. The presence of vinyl versatate (derived from a synthetic C_{10} branched chain saturated fatty acid) is indicated by ions at m/z 102, 87, and 88 while m/z 127 is characteristic of the melamine modification (Figure 8A). When rosin is incorporated into an alkyd (Figure 8B), high mass ions at m/z 239 and 197 result from the pyrolysis of one of the constituents, i.e., dehydroabietic acid. Figure 8C shows the very characteristic spectrum of a urethane-modified long oil alkyd in which the ion at m/z 174 is due to toluene di-isocyanate. The presence of styrene (Figure 9A) can be deduced by the increase in the ratio of the intensities of ions m/z 104 to that at m/z 76, and the presence of an ion of significant intensity at m/z 103. Occasionally, alkyds based on iso-phthalic acid occur and this is readily seen (Figure 9B) by the intense ion at m/z 105 with others at m/z 122 and 77 due to benzoic acid. These ions can also be seen but to a much lesser extent in the normal alkyd-based phthalic anhydride when benzoic acid is employed as a "chain-stopper". Vinyl toluene modification leads to highly characteristic ions (Figure 9C) of vinyl toluene at m/z 118, 117, 115, and 91 with the ion at m/z 105 probably arising from ethyl benzene.

Modifications to alkyds involving nitrocellulose, urea, or silicones are not readily detected except at high levels using flash pyrolysis methods. In the cases of nitrocellulose and urea modifications, this is mainly due to the fact that their pyrolysis products are at low mass and are masked by ions which arise from the bulk of the resin. In the case of silicone modification, the levels are less than 1%, which is too low for their

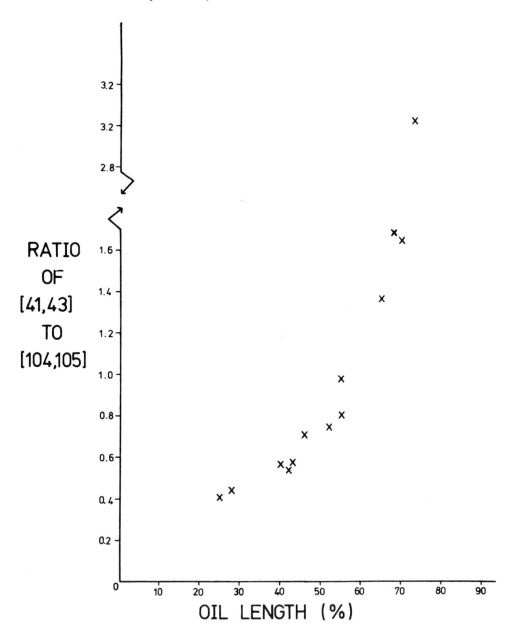

FIGURE 7.   Correlation of ion intensity ratios with alkyd resin "oil lengths".

reliable detection since the characteristic ion at m/z 207 can also arise by contamination with traces of silicone oils. However, it is possible to detect the presence of nitrocellulose in alkyds by the use of a slow thermally ramped probe.

The TIC trace for a nitrocellulose-alkyd blend shown in Figure 10A was obtained by programming the probe from ambient to 650 at 60°C/min. It shows that the thermal decomposition proceeds by different routes as the temperature is ramped. In the early phase (up to scan 35) the main products are phthalate plasticizers. The spectrum taken at scan 41 (Figure 10B) is consistent with the products formed from the pyrolysis of nitrocellulose, i.e., N_2 (m/z 28), NO (m/z 30), CO_2 (m/z 44), and NO_2 (m/z 46). The final peak in the TIC trace corresponds to the degradation of the alkyd resin itself;

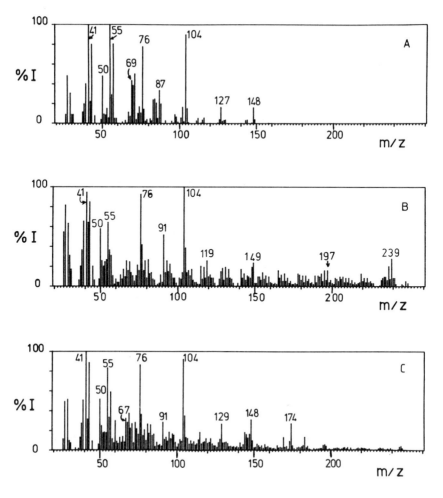

FIGURE 8.   Mass pyrograms of alkyd resins modified with (A) vinyl versatate and melamine, (B) rosin, and (C) urethane.

Figure 10C shows the spectrum obtained by averaging the spectra over scans 35 to 80. Mass chromatograms for NO (m/z 30) and phthalic anhydride (m/z 104) can also be used to indicate the degree of modification of the resin. Pidduck[85] has used this method to study a range of dried alkyd binders, and the overall spectra are in reasonable agreement with the results obtained in this laboratory for similar resins. Ardrey et al.[86] have used the VG Pyrolysis Probe to study ten white alkyd-based paints. Of these, two were modified with vinyl toluene and toluene di-isocyanate, respectively, while the remaining ones could be differentiated to a moderate degree on the basis of 20 significant ions. A factor that can be used for the discrimination of alkyds using pyrolysis/GC is the acrolein-to-methacrolein ratios,[87] these compounds being pyrolysis products of the polyols glycerol and pentaerythritol, respectively. Although ions in the mass pyrograms at m/z 56 (acrolein) and m/z 70 (methacrolein) could be seen[86] to vary from sample to sample, there was also considerable variation between duplicates. It is obvious that other products are contributing to the intensity of the ions at these masses, since it was shown by pyrolysis/GC[87] that acrolein and methacrolein are formed reproducibly in the pyrolysis of alkyds. A different approach to the differentiation of alkyds was taken by Smalldon et al.,[44] who used thermogravimetry (TG)/MS to study ten white paints. They obtained TIC traces which allowed these paints to be classified into

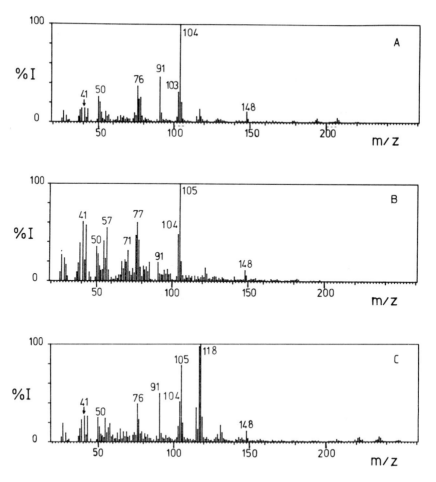

FIGURE 9.   Mass pyrograms of alkyd resins modified with (A) styrene, (B) iso-phthalic acid, and (C) vinyl toluene.

seven groups. A disadvantage of this method, however, is that large (100-μg) samples are required, although it is claimed that only 10 μg is necessary if only characteristic ions are monitored. However, in practice this could not be used for real forensic samples for which the quantity of sample is limited.

Thermosetting acrylics produce a variety of monomers on pyrolysis, methyl methacrylate, butyl acrylate, and ethylhexyl acrylate being the ones most frequently encountered. Nitrogen-containing resins such as melamine are also often incorporated to promote cross-linking, and frequently styrene is added to enhance the drying properties. Variations in formulation create a wide diversity of pyrograms and since acrylics pyrolyze very much more reproducibly than alkyds, high discrimination is readily achieved. An example (Figure 11), taken from casework involving a two-way transfer of paint between vehicles in collision, illustrates the type of spectra obtained from such resins and the ease with which they may be differentiated. Yellow paint (Figure 11A) found on the red Alpine car was a styrenated acrylic based on methyl methacrylate and matched the control from the yellow Escort car. Similarly, the red paint (Figure 11B) found on the latter matched that of the Alpine and was essentially a styrenated butyl methacrylate-based resin. As has been mentioned before, although the mass pyrograms at 70 eV of these resins may be readily differentiated on the basis of either unique ions or comparison of relative intensities, the use of CI brings out the differences even more

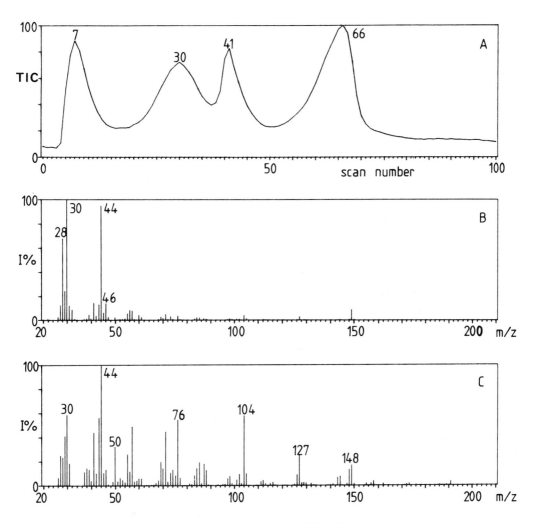

FIGURE 10. Ramped pyrolysis/MS of a nitrocellulose-modified alkyd resin: (A) total ion current trace, (B) scan 41, and (C) averaged mass spectrum (scans 35 to 80).

dramatically (see Figure 12). Saferstein and Manura[29] have shown that pyrolysis/MS under CI conditions can be used to discriminate automobile paints of this type. The interpretation of the pyrograms is not without ambiguity due to the number of isomeric species which could be attributed to the quasimolecular ion and their tendency to fragment, usually by the loss of H_2O. Nevertheless, the presence of (MH)⁺ ions corresponding to methyl methacrylate (m/z 101), styrene (m/z 105), butyl acrylate (m/z 129), and butyl methacrylate (m/z 143) were readily discerned from the spectra.

Emulsion paints are relatively unimportant forensically and, in almost all cases encountered, are based on polyvinyl acetate resins frequently copolymerized with vinyl versatate. The mass pyrogram of a typical emulsion paint, shown in Figure 13, exhibits ions characteristic of acetic acid (m/z 43 and 60), toluene (m/z 91 and 92), benzene (m/z 78), and naphthalene (m/z 128) arising essentially from the degradation of polyvinyl acetate. The presence of vinyl versatate can be seen by the relatively low intensity, but significant ions are at m/z 87, 88, and 102.

In an attempt to assess the overall discriminatory potential of pyrolysis/MS for forensic paint analysis, we have surveyed the results obtained in the three most common case sample categories. Vehicle paints were shown to be approximately two thirds alk-

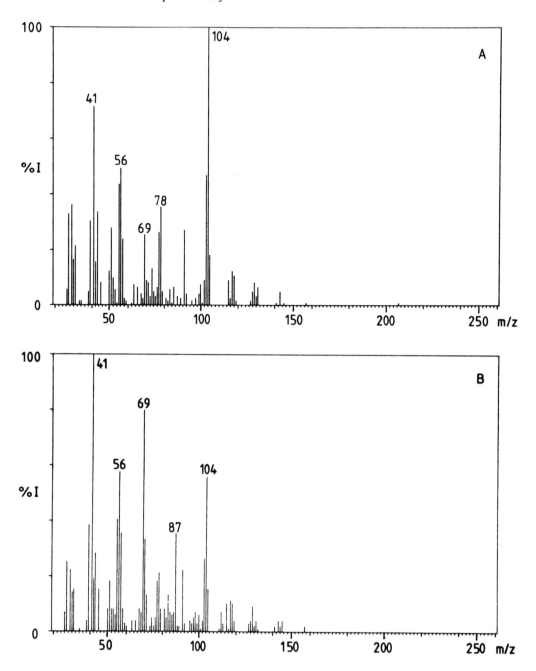

FIGURE 11.   Mass pyrograms of paint samples involving a two-way transfer during a vehicle collision: (A) paint from a yellow Ford Escort and (B) paint from a red Chrysler Alpine.

yds with the rest being mainly acrylics. The alkyds could be further subdivided in terms of modification with melamine and on the basis of oil length. Further discrimination could also be achieved on the basis of a comparison of the particular ion intensities. In the case of decorative gloss paints, a survey showed that the majority of paints were unmodified alkyds with those of medium oil lengths predominating. However, a pair-wise comparison of the cited selected ions could achieve an overall discrimination of 83%. Tool paints, in contrast, were composed of a wider range of resins although

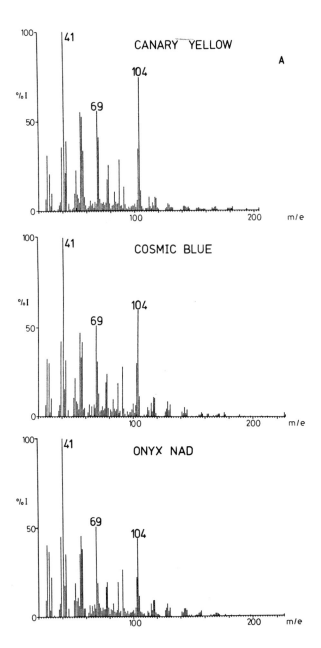

FIGURE 12.   Mass pyrograms obtained from three acrylic paints using (A) 70 eV EI and (B) isobutane CI.

alkyds predominate. Of the 52 samples examined, 27 exhibited unique spectra with the typical alkyd group being capable of further discrimination in a similar fashion to that described for the decorative glosses.

Paint pyrograms, as has been seen, display a considerable diversity and if information as to the formulation of the paints were more readily available, they could be used for more than just comparative purposes. For example, paint formulations are changed for various reasons and, in principle, careful monitoring might permit an unknown sample to be assigned to a particular production year. This type of information could on occasion have considerable forensic significance.

B

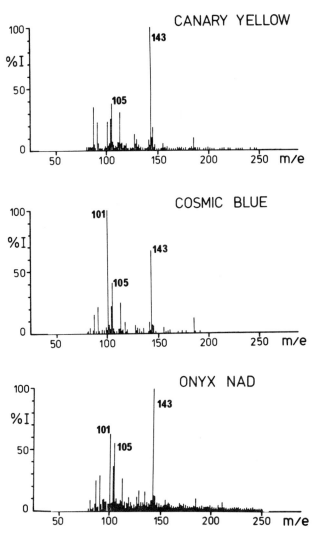

FIGURE 12B.

2. Adhesives

This type of material arises in a variety of situations, most commonly from cases involving the alteration of documents or the resealing of envelopes and parcels. They are also frequently encountered in homemade explosive devices and occasionally acts of vandalism, e.g., the jamming of locks with epoxy adhesives, etc. The utility of pyrolysis/MS for the examination of these materials was first demonstrated in this laboratory[88] and since that time the collection of adhesives readily available for do-it-yourself (DIY) purposes has been updated. In a recent survey by pyrolysis/MS of approximately 100 manufactured adhesive products,[89] it was shown that they can be readily identified as belonging to the class of compound shown in the Table 4, and considerable discrimination within class is often possible. Most of these types of adhesives have been encountered in casework over the past 10 years with the exception of a twin-

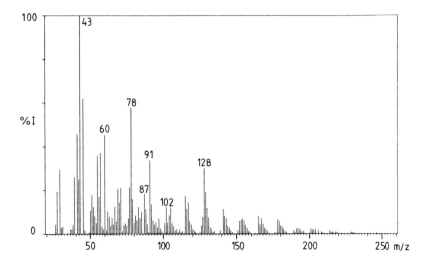

FIGURE 13. Mass pyrogram of a vinyl acetate/vinyl versatate-based emulsion paint.

Table 4
CLASSIFICATION OF ADHESIVES BY
CHEMICAL TYPE[a]

Acrylics	Polystyrenes
Carbohydrates	Polyurethanes
Cyanoacrylates	Polyvinylacetate/polyvinylalcohol
Epoxies	Proteins
Natural rubbers	Silicone rubbers
Neoprene rubbers	Styrene-butadiene rubbers
Nitrile rubbers	

[a]   Based on a survey in 1985 at the Metropolitan Police
Forensic Science Laboratory.

pack acrylic which consists of tetrahydrofurfuryl methacrylate and a diphenylamine hardener. The mass pyrograms of four rubber-like adhesives are shown in Figure 14, from which it can be seen that all show characteristic spectra.

The use of a direct-insertion probe (programmed from 30 to 500°C at 60°C/min) for the examination of polyurethane and epoxy resins used as sealants has been described.[37] Selected ion profiling was found to be a convenient way of distinguishing between 12 or so such materials that were routinely encountered. For example, Figure 15 shows the total ion and selected ion profiles for two epoxy resins of similar composition. Although the TIC profiles are very similar, they differ in respect to their phenol (m/z 94) and alkyl phenol (m/z 107) content.

At the FBI laboratories, 91 commercially available adhesives were studied by pyrolysis-capillary column GC/MS.[4] A composite spectrum (Figure 16A) was generated by computer summation of over 1500 spectra from each pyrogram (Figure 16B). It was claimed that the composite spectrum reflected all the components and their relative concentrations in each pyrogram. In this example, the presence of butadiene (m/z 39), benzene (m/z 78), toluene (m/z 91 and 92), styrene (m/z 104 and 78), indene (m/z 115 and 116), methyl styrene (m/z 115, 117, and 118), and naphthalene (m/z 128) can all be readily seen. The advantage of this approach is that the spectra can be readily interpreted and a sample library built up for comparison purposes. The disadvantage lies in

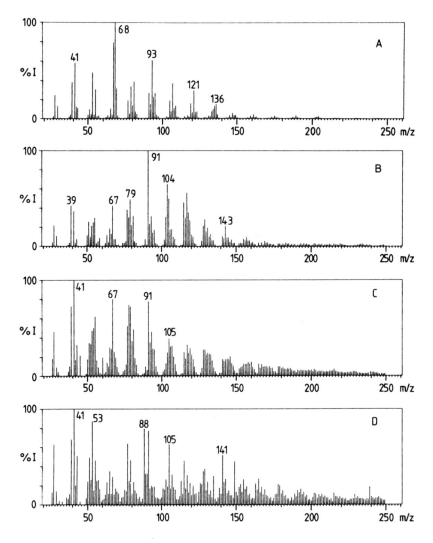

FIGURE 14.   Mass pyrograms of (A) polyisoprene, (B) styrene-butadiene, (C) nitrile, and (D) neoprene-based rubber-like adhesives.

the longer analysis time, particularly when a large number of comparisons must be made.

3. Rubbers and Foams

Synthetic polymers are extensively used for the soles and heels of shoes, and smears from footwear are occasionally submitted for analysis. Nitrile, styrene/butadiene, and polyisoprene rubbers have all been encountered in casework and, although each class can be readily distinguished, individual samples show similar spectra to rubber-like adhesives. However, the use of the polymer may be adduced easily by its appearance since rubbers are invariably filled with carbon black whereas adhesives of this type are usually transparent.

A limited survey has also been carried out by pyrolysis/MS of tire sidewall and tread rubbers. The uniformity of composition (23% styrene, 77% butadiene) of tread rubbers was reflected in the very similar mass pyrograms, an example of which is shown in Figure 17A. Sidewall rubbers, on the other hand, showed differences attributable to

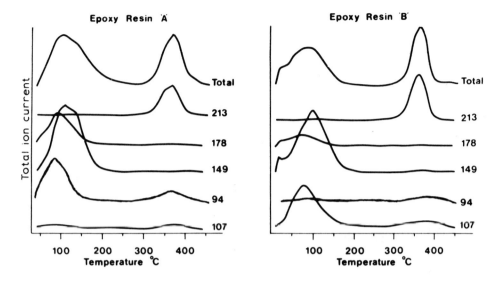

FIGURE 15. Mass pyrograms for epoxy resin (m/z 213), pitch (m/z 178), dialkyl phthalate (m/z 149), alkyl phenols (m/z 107), and phenol (m/z 94). (From Williamson, J. E., Cocksedge, M. J., and Evans, N., *J. Anal. Appl. Pyrol.*, 7B, 1473, 1978. With permission.)

a variation in their styrene/butadiene to natural rubber content. The differences are manifested in the intensities of ions at m/z 41, 67, 68, 91, 104, and 105 (c.f. Figures 17B and C). On the basis of this limited amount of work it would appear that pyrolysis/MS would be capable of characterizing rubber smears from the sidewalls of tires in the event of vehicle incidents.

Synthetic foams, such as those used in upholstery and as draught excluders, are polyurethanes based on toluene di-isocyanate (TDI) or diphenyl methane di-isocyanate (MDI) and a variety of polyglycols or polyethers. These arise from time to time in casework and show characteristic mass pyrograms, examples of which are shown in Figure 18, the ions at m/z 174 and 250 and 224 being characteristic of TDI and MDI, respectively. A wide variety of polyurethanes used for various purposes, including foams, have been studied by Marshall,[39] who used high-temperature probe MS.

4. Fibers

An extensive study[90] of fibers was conducted in this laboratory during the early phase of our pyrolysis/MS research and it was found that rapid characterization of short (0.5- to 1.0-mm) single fibers could be achieved. This study included nylons, polyesters, acrylics, polyolefins, chlorine-containing polymers, cellulosics, and various natural fibers. Figure 19 shows some results obtained for nylons from which it can be seen that they can all be readily distinguished. Unfortunately, the results obtained at that time on the forensically important class of acrylic fibers was less satisfactory. Although some discrimination was obtained with the extensively modified acrylics, the technique was not capable of the degree of differentiation afforded by IR spectroscopy for those examples with only minor modifications.

Saferstein and Manura[29] have demonstrated that pyrolysis/MS using isobutane CI may be employed for characterizing acrylic fibers. However, only three fibers were examined and no attempt was made to interpret the results obtained. Ardrey et al.[86] have used the VG Pyrolysis Probe and thermogravimetry/MS[91] to study acrylic fibers. They concluded that pyrolysis/MS and thermogravimetry/MS were complementary, the former allowing rapid characterization of the polymer type while the latter, although a more time-consuming technique, allowed a greater degree of discrimination

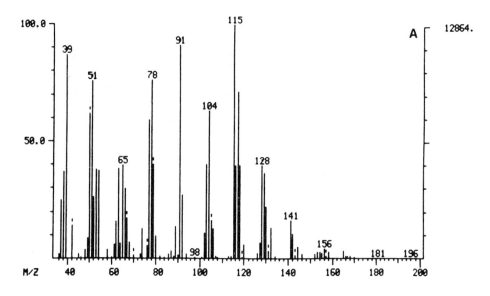

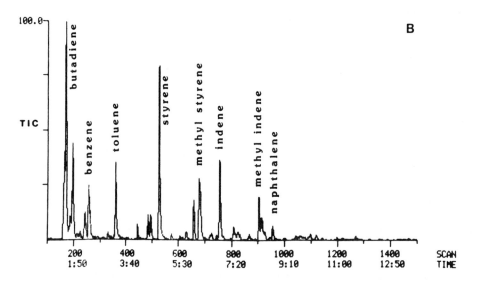

FIGURE 16. (A) Composite mass spectrum derived from pyrolysis/GC/MS of an adhesive; (B) total ion current trace of same adhesive. (From Bakowski, N. L., Bender, E. C., and Munson, T. O., *J. Anal. Appl. Pyrol.,* 8, 483, 1985. With permission.)

to be achieved. A similar study by this group[25] of polyesters concluded that pyrolysis/MS could distinguish readily between chemically distinct polyesters, but those based on polyethylene terephthalate gave qualitatively similar spectra.

At this moment it would appear that pyrolysis/MS will only have a limited role to play in the forensic examination of fibers since the technique does not have the overall discrimination that has been demonstrated for IR spectroscopy. Furthermore, developments in sampling methods and combination with Fourier transform infrared spectroscopy[92] makes IR a more appealing technique, especially as it is nondestructive.

5. Structural Plastics

Despite the extensive use of plastics for decorative and structural purposes in the modern home and workplace, such samples are not submitted with any frequency since

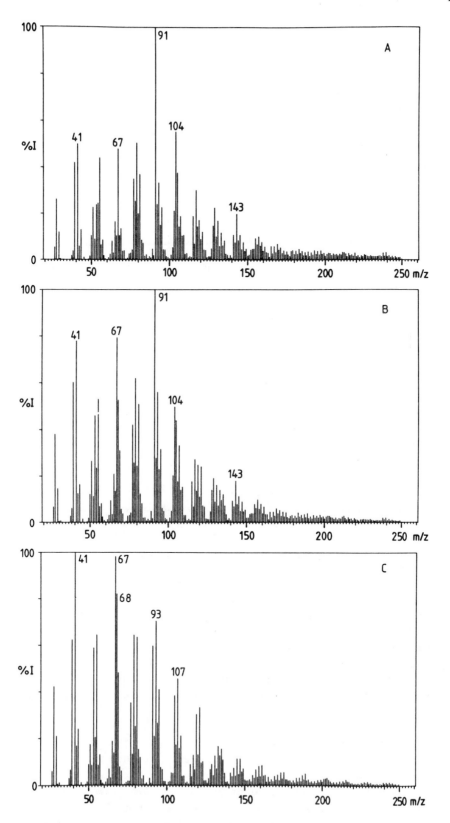

FIGURE 17.   Mass pyrograms from (A) a Michelin ZX tire tread, (B) a Dunlop tire side-wall, and (C) an Avon tire sidewall.

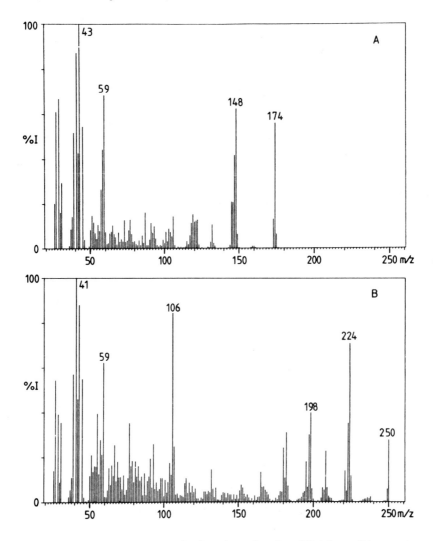

FIGURE 18.   Mass pyrograms of polyurethanes based on (A) toluene di-isocyanate and (B) diphenyl methane di-isocyanate.

there is less likelihood of their transfer than is the case with, e.g., paints. With vehicles, however, where an increasing use of plastics is being made, the likelihood of transfer during collision is high. Over the last few years, a diverse range of polymers have been characterized by pyrolysis/MS in these cases. These have included polyethylene, polypropylene, polyvinylchloride and its copolymers with styrene and acrylates, polyvinyltoluene copolymers, polymethylmethacrylate, and polyurethanes based on diphenyl di-isocyanate. The mass pyrograms shown in Figure 20 illustrate the range of materials which can be encountered in just such a case.

6. Wire Coverings and Adhesive Tapes

The highly plasticized PVC covering of many electrical wires is one type of sample which generally gives a rather uninformative mass pyrogram. The plasticizer (typically a di-alkyl phthalate) dominates the initial spectra taken soon after the insertion of the Pyroprobe in the expansion chamber. On pyrolysis of the PVC, HCl is the major product evolved, but in our system it sometimes binds to the expansion chamber or the inlet line, and may therefore make only a small contribution to the overall mass pyro-

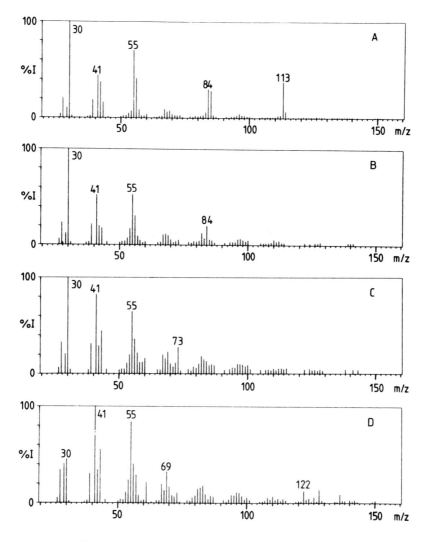

FIGURE 19. Mass pyrograms of (A) Nylon 6, (B) Nylon 66, (C) Nylon 6:10, and (D) Nylon 11.

gram. The major ions usually observed in these circumstances are attributable to benzene, toluene, and naphthalene. In general, little differentiation may be made between different PVCs due to the somewhat irreproducible nature of the mass pyrograms.

Plastic-backed adhesive tapes are frequently made from PVC and the same limitations apply in their analysis by pyrolysis/MS. However, examination of the adhesive layer can be more informative as the example of PVC tape (Figure 21A and B) with a nitrile rubber adhesive shows.

7. Type Correcting Fluids

The examination of documents which contain alterations involving type correcting fluids has been studied by pyrolysis/MS.[93] The polymeric base in such formulations has been found to show considerable variation between manufacturers, and in a study involving ten different products each pyrogram was unique, although acrylates, vinyl toluene, and styrene-containing mixtures were indicated. A typical case example of this nature involved the type correcting fluid Tipp-Ex®, the mass pyrogram of which is shown in Figure 22 and may be attributed to *sec*-butyl methacrylate.

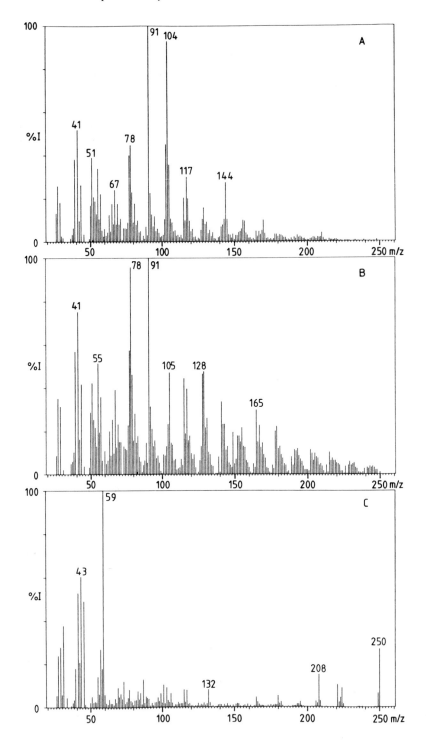

FIGURE 20. Mass pyrograms of plastics from a car interior (A) dashboard upper fascia (ABS), (B) glove compartment keyhole surround (PVC), and (C) steering wheel (diphenyl methane di-isocyanate polyurethane).

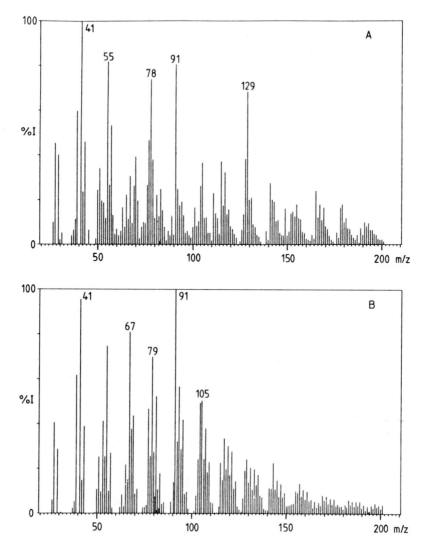

FIGURE 21.   Mass pyrograms of an adhesive tape: (A) backing (PVC) and (B) adhesive layer (nitrile rubber).

V. CONCLUDING REMARKS

From the examples given in Section IV, it can be seen that pyrolysis/MS can provide valuable information on synthetic polymers both for general analytical purposes and in a forensic context. It has the advantages of speed and allows for large databases to be readily created and the technique could, if required, be automated. However, its main drawbacks are that expensive instrumentation is required and in some instances the interpretation of the data can be ambiguous. There is a need for additional work to establish the optimum instrumental conditions for a particular analytical purpose and also to determine its reproducibility. The assessment of pyrolysis/MS data, certainly from the forensic point of view, is somewhat primitive at the moment and more effort is required to evaluate the sophisticated methods that have been used in other fields.

It is to be hoped, however, that this chapter has alerted others to the potential of

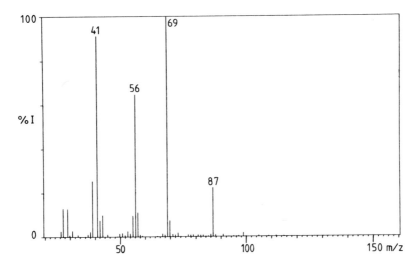

FIGURE 22. Mass pyrogram of Tipp-Ex® type correcting fluid.

pyrolysis/MS and overcome some of the mystique and confusion that has grown up about the technique in the general scientific literature. Certainly the authors would welcome more participants in the field, both forensic scientists and those from the scientific community in general. Most forensic science laboratories in Europe and North America have mass spectrometers, usually employed for GC/MS purposes, which could be readily adapted for pyrolysis/MS without prejudicing their primary function. By attracting more people to the field, wider experience and additional information will be obtained about the technique. In this way, more rapid progress will be made in overcoming the problems already alluded to, and eventually the technique will become fully accepted and established.

ACKNOWLEDGMENTS

The authors wish to thank John Hughes, Ian Jane, Dave Hickman, Chris Curry, and Phil Burke for the valuable contribution they have made to the development of pyrolysis/MS in the Metropolitan Police Forensic Science Laboratory over the last 10 years. We would also like to thank Barbara Whitehouse for her patient and painstaking assistance with the preparation of the manuscript.

REFERENCES

1. Foltz, R. L., Neher, M. B., and Hinnenkamp, E. R., Applications of mass spectrometry and gas chromatography to the analysis of polymer systems, *Appl. Polym. Symp.*, 10, 195, 1969.
2. Wuepper, J. L., Pyrolysis gas chromatographic-mass spectrometric identification of intractable materials, *Anal. Chem.*, 51, 997, 1979.
3. Alajbeg, A., Arpino, P., Deur-Siftar, D., and Guichon, G., Investigation of some vinyl polymers by pyrolysis gas chromatography-mass spectrometry, *J. Anal. Appl. Pyrol.*, 1, 203, 1980.
4. Bakowski, N. L., Bender, E. C., and Munson, T. O., Comparison and identification of adhesives used in improvised explosive devices by capillary column gas chromatography-mass spectrometry, *J. Anal. Appl. Pyrol.*, 8, 483, 1985.
5. Zemany, P. D., Identification of complex materials by mass spectrometric analysis of their pyrolysis products, *Anal. Chem.*, 24, 1709, 1952.

6. Meuzelaar, H. L. C. and Kistemaker, P. G., A technique for fast and reproducible fingerprinting of bacteria by Py-MS, *Anal. Chem.*, 45, 587, 1973.

7. Irwin, W. J., Analytical pyrolysis — an overview, *J. Anal. Appl. Pyrol.*, 1, 3, 1979.

8. Irwin, W. J., Bibliography, *J. Anal. Appl. Pyrol.*, 3, 3, 1981.

9. Haverkamp, J. and Kistemaker, P. G., Recent developments in pyrolysis mass spectrometry, *Int. J. Mass Spectrom. Ion Phys.*, 45, 275, 1982.

10. Meuzelaar, H. L. C., Windig, W., Harper, A. M., Huff, S. M., McClennen, W. H., and Richards, J. M., Pyrolysis mass spectrometry of complex organic materials, *Science*, 226, 268, 1984.

11. Schulten, H.-R. and Lattimer, R. P., Applications of mass spectrometry to polymers, *Mass Spectrom. Rev.*, 3, 231, 1984.

12. Meuzelaar, H. L. C., Haverkamp, J., and Hileman, F. D., *Pyrolysis Mass Spectrometry of Recent and Fossil Biomaterials*, Elsevier, Amsterdam, 1982.

13. Irwin, W. J., *Analytical Pyrolysis. A Comprehensive Guide*, Marcel Dekker, New York, 1982.

14. Vorhees, K. J., Ed., *Analytical Pyrolysis. Techniques and Applications*, Butterworths, London, 1984.

15. Schulten, H.-R., Ed., Proceedings of the Sixth International Symposium on Analytical and Applied Pyrolysis, Wiesbaden, 1984, *J. Anal. Appl. Pyrol.*, 8, 1985.

16. Wheals, B. B., Analytical pyrolysis techniques in forensic science, *J. Anal. Appl. Pyrol.*, 2, 277, 1984.

17. Wheals, B. B., The practical application of pyrolytic methods in forensic science during the last decade, *J. Anal. Appl. Pyrol.*, 8, 503, 1985.

18. Saferstein, R., Forensic aspects of analytical pyrolysis, *Chromatogr. Sci.*, 29, 339, 1985.

19. Meuzelaar, H. L. C. and Huff, S. M., Characterisation of leukemic and normal cells by Curie-point pyrolysis-mass spectrometry, *J. Anal. Appl. Pyrol.*, 3, 111, 1981.

20. Shute, L. A., Gutteridge, C. S., Norris, J. R., and Berkley, R. C. W., Curie-point pyrolysis mass spectrometry applied to characterisation of selected bacillus species, *J. Gen. Microbiol.*, 130, 343, 1984.

21. Schmid, P. P. and Simon, W., A technique for Curie-point mass spectrometry with a Knudsen reactor, *Anal. Chim. Acta*, 89, 1, 1977.

22. Hughes, J. C., Wheals, B. B., and Whitehouse, M. J., Simple technique for pyrolysis-mass spectrometry of polymeric materials, *Analyst*, 102, 143, 1977.

23. Bracewell, J. M. and Robertson, G. W., Characteristics of soil organic matter in temperate soils. Curie-point pyrolysis-mass spectrometry. I. Organic matter variations with drainage and mull humification in an A horizon, *J. Soil Sci.*, 35, 549, 1984.

24. Whitehouse, M. J., Metropolitan Police Forensic Science Laboratory, unpublished results.

25. Ardrey, R. E., Batchelor, T. M., and Smalldon, K. W., A probe for pyrolysis-mass spectrometry and its application to the characterisation of commercial polyester fibres, Report No. 311, Home Office Central Research Establishment, 1979.

26. Hickman, D. A. and Jane, I., Reproducibility of pyrolysis-mass spectrometry using three different pyrolysis systems, *Analyst*, 104, 334, 1979.

27. Chemical Data Systems, Inc., Oxford, Pa.

28. Wells, G., Futtrell, J. H., and Vorhecs, K. J., Pyrolysis-mass spectrometry for direct sampling of primary products of thermolysis reactions, *Rev. Sci. Instrument.*, 52, 735, 1981.

29. Saferstein, R. and Manura, J., Pyrolysis mass spectrometry — a new forensic science technique, *J. Forensic Sci.*, 22, 748, 1977.

30. Udseth, H. R. and Friedman, L., Analysis of styrene polymers by mass spectrometry with filament heated evaporation, *Anal. Chem.*, 53, 29, 1981.

31. Israel, S. C., Bechard, M. J., and Abbot, M., Polymer characterisation by direct pyrolysis chemical ionisation mass spectrometry, *Polym. Prep.*, 24, 159, 1983.

32. Risby, T. H., Yergey, J. A., and Scocca, J. J., Linear programmed thermal degradation mass spectroscopy of polystyrene and polyvinylchloride, *Anal. Chem.*, 54, 2228, 1982.

33. Zeman, A., The identification of some commercial polymers by thermal decomposition in the mass spectrometer, *Angew. Makromol. Chem.*, 31, 1, 1973.

34. Ballistreri, A., Foti, S., Montaudo, G., Pappalardo, S., Scamporrino, E., Arnesano, A., and Calgari, S., Thermal decomposition of acrylonitrile copolymers investigated by direct pyrolysis-mass spectrometry, *Makromol. Chem.*, 180, 2835, 1979.

35. Mischer, G., Mass spectrometry of high polymers, *Adv. Mass Spectrom.*, 78, 1444, 1978.

36. Luderwald, I., Przybylski, M., and Ringsdorf, H., Pyrolysis of polymers in the mass spectrometer, *Adv. Mass Spectrom.*, 7B, 1473, 1978.

37. Williamson, J. E., Cocksedge, M. J., and Evans, N., Analysis of polyurethanes and epoxy resin based materials by pyrolysis-mass spectrometry, *J. Anal. Appl. Pyrol.*, 2, 195, 1980.

38. Marshall, G. L., Pyrolysis mass spectrometry of polymers. I. Unsaturated polyesters based on maleic anhydride, *Eur. Polym. J.,* 18, 53, 1982.

39. Marshall, G. L., Pyrolysis mass spectrometry of polymers. II. Polyurethanes, *Eur. Polym. J.,* 19, 439, 1983.

40. Pidduck, A. J., Mass spectrometric analysis of halogenated polymers, *J. Anal. Appl. Pyrol.,* 7, 215, 1985.

41. Curry, C. J. and Whitehouse, M. J., Metropolitan Police Forensic Science Laboratory, to be published.

42. Mol, G. J., Gritter, R. J., and Adams, G. E., Mass spectrometry of thermally treated polymers, in *Applied Polymer Spectroscopy,* Academic Press, New York, 1978, 257.

43. Morisaki, S., Simultaneous thermogravimetry-ms and pyrolysis gas chromatography of fluorocarbon polymers, *Thermochim. Acta,* 25, 171, 1978.

44. Smalldon, K. W., Ardrey, R. E., and Mullings, L. R., The characterisation of closely related polymers by thermogravimetry-ms, *Anal. Chim. Acta,* 107, 327, 1979.

45. Carracher, C. E., Jr., Molloy, H. M., Taylor, M. L., Tiernan, T. O., Yelton, R. O., Schroeder, T. A., and Bogdan, M. R., Identification of degradation products and structures of organometallic polymers through coupled thermogravimetric-ms, *Org. Coat. Plast. Chem.,* 41, 197, 1976.

46. Kistemaker, P. G., Boerboom, A. J. H., and Meuzelaar, H. L. C., Laser pyrolysis mass spectrometry and applications to technical polymers, *Dyn. Mass Spectrom.,* 4, 139, 1976.

47. Lum, R. M., Direct analysis of polymer pyrolysis using laser microprobe techniques, *Thermochim. Acta,* 18, 73, 1977.

48. Unsold, E., Hillenkamp, F., Renner, G., and Nitsche, R., Investigations of organic materials using LAMMA, *Adv. Mass Spectrom.,* 7B, 1425, 1978.

49. Hercules, D. M., Solid state mass spectrometry using a laser microprobe, in *Analytical Pyrolysis,* Vorhees, K. J., Ed., Butterworths, London, 1984, 1.

50. LAMMA 1000 analysis of polymers, technical note, Leybold-Heraeus GmbH.

51. Windig, W., Kistemaker, P. G., Haverkamp, J., and Meuzelaar, H. L. C., The effects of sample preparation, pyrolysis and pyrolysate transfer conditions of Py-MS, *J. Anal. Appl. Pyrol.,* 1, 39, 1979.

52. Meuzelaar, H. L. C., Kistemaker, P. G., Eshuis, W., and Boerboom, A. J. H., Automated Py-Ms: application to the differentiation of micro-organisms, *Adv. Mass Spectrom.,* 7B, 1452, 1978.

53. Aries, R. E., Gutteridge, C. S., and Ottley, T. W., Evaluation of a low-cost automated pyrolysis mass spectrometer, to be published in J. Anal. Appl. Pyrol.

54. McLafferty, F. W., *Interpretation of Mass Spectra,* 2nd ed., W. A. Benjamin, New York, 1973.

55. Meuzelaar, H. L. C., Posthumus, M. A., Kistemaker, P. G., and Kistemaker, J., Curie-point pyrolysis in direct combination with low voltage electron impact ionization mass spectrometry, *Anal. Chem.,* 45, 1546, 1973.

56. Meuzelaar, H. L. C., Haverkamp, J., and Hileman, F. D., *Pyrolysis Mass Spectrometry of Recent and Fossil Biomaterials,* Elsevier, Amsterdam, 1982, 52.

57. Millard, B. J., *Quantitative Mass Spectrometry,* Heydon, London, 1979, 45.

58. Whitehouse, M. J., Boon, J. J., Bracewell, J. M., Gutteridge, C. S., Pidduck, A. J., and Puckey, D. J., Results of a pyrolysis mass spectrometry interlaboratory trial, *J. Anal. Appl. Pyrol.,* 8, 515, 1985.

59. Chatfield, D. A., Hileman, F. D., Vorhees, K. J., and Futrell, J. H., Characterisation of polymer decomposition products by electron impact and chemical ionization mass spectrometry, in *Applications of Polymer Spectroscopy,* Academic Press, New York, 1978, 241.

60. Shimizi, Y. and Muson, B., Pyrolysis/chemical ionization mass spectrometry of polymers, *J. Polym. Sci. Polym. Chem. Educ.,* 17, 1991, 1979.

61. Adams, R. E., Pyrolysis mass spectrometry of terephthalate polyesters using negative ionization, *J. Polym. Sci. Polym. Chem. Educ.,* 20, 119, 1982.

62. Adams, R. E., Positive and negative chemical ionization pyrolysis mass spectrometry of polymers, *Anal. Chem.,* 55, 414, 1983.

63. Conway, D. C. and Marak, R., Analysis of polymers by pyrolysis chemical ionization mass spectrometry, *J. Polym. Sci. Polym. Chem. Educ.,* 20, 1765, 1982.

64. Schuddemage, H. D. R. and Hummel, D. O., Characterisation of polymers by pyrolysis with field ionization mass spectrometry, *Adv. Mass Spectrom.,* 4, 857, 1968.

65. Hummel, D. O., Structure and degradation behaviour of synthetic polymers in combination with FIMS, in *Analytical Pyrolysis,* Jones, C. E. R. and Cramers, C. A., Eds., Elsevier, Amsterdam, 1977, 117.

66. Gortz, W., Schulten, H.-R., and Beckey, H. D., FI- and FD-MS in analytical chemistry, in *Mass Spectrometry,* Part A, Merritt, C., Jr. and McEwen, C. N., Eds., Marcel Dekker, New York, 1979, 145.

67. Bahr, U., Luederwald, I., Meuller, R., and Schulten, H.-R., Pyrolysis field desorption mass spectrometry. III. Aliphatic amides, *Angew. Makromol. Chem.,* 120, 163, 1984.
68. Doerr, M., Luederwald, I., and Schulten, H.-R., Characterisation of polymers by field desorption and fast atom bombardment, *Fresenius Z. Anal. Chem.,* 318, 339, 1984.
69. Doerr, M., Luederwald, I., and Schulten, H.-R., Investigations of polymers by field desorption and fast atom bombardment mass spectrometry, *J. Anal. Appl. Pyrol.,* 8, 109, 1985.
70. Genui, W. and Boon, J. J., Pyrolysis gas chromatography photo-ionisation mass spectroscopy, a new approach in the analysis of macromolecular materials, *J. Anal. Appl. Pyrol.,* 8, 25, 1985.
71. Meuzelaar, H. L. C., Haverkamp, J., and Hileman, F. D., *Pyrolysis Mass Spectrometry of Recent and Fossil Biomaterials,* Elsevier, Amsterdam, 1982.
72. Levsen, K. and Schulten, H.-R., Analysis of mixtures by CAMS: pyrolysis products of DNA, *Biomed. Mass Spectrom.,* 3, 137, 1976.
73. Foti, S., Liguori, A., Maravigna, P., and Montaudo, G., Characterisation of poly-(carboxypiperazine) by mass analysed ion kinetic energy spectrometry, *Anal. Chem.,* 54, 674, 1982.
74. Holtzman, G. and Kossmehl, G., Application of linked scans in pyrolysis mass spectroscopy of polymers, *Biomed. Mass Spectrom.,* 15, 336, 1980.
75. Dawson, P. H., Ed., *Quadrupole Mass Spectrometry and its Applications,* Elsevier, Amsterdam, 1976.
76. Meuzelaar, H. L. C., Haverkamp, J., and Hileman, F. D., *Pyrolysis Mass Spectrometry of Recent and Fossil Biomaterials,* Elsevier, Amsterdam, 1982, 53.
77. Tuithof, H. H., Boerboom, A. J. H., Kistemaker, P. G., and Meuzelaar, H. L. C., A magnetic mass spectrometer with simultaneous ion detection and variable mass dispersion in laser pyrolysis and collision induced dissociation studies, *Adv. Mass Spectrom.,* 7B, 838, 1978.
78. Chapman, J. R., *Computers in Mass Spectrometry,* Academic Press, New York, 1978.
79. Nie, N. H., Hull, C. H., Jenkins, J. G., Steinbrenner, K., and Bent, D. H., *Statistical Package for the Social Sciences,* 2nd ed., McGraw-Hill, New York, 1975.
80. Petraco, N., The occurrence of trace evidence in one examiner's casework, *J. Forensic Sci.,* 30, 455, 1985.
81. Hickman, D. A., A classification scheme for glass, *Forensic Sci. Int.,* 17, 265, 1981.
82. Windig, W., Haverkamp, J., and Kistemaker, P., Interpretation of sets of pyrolysis mass spectrometry spectra by discriminant analysis and graphical rotation, *Anal. Chem.,* 55, 81, 1983.
83. Windig, W. and Meuzelaar, H. L. C., Nonsuperimposed numerical component extraction from pyrolysis mass spectra of complex mixtures, *Anal. Chem.,* 56, 2297, 1984.
84. Vallis, L. H., MacFie, H. J., and Gutteridge, C. S., Comparison of cannonical variate analysis with target rotation and least square regression as applied to pyrolysis mass spectra of simple biochemical mixtures, *Anal. Chem.,* 57, 704, 1985.
85. Pidduck, A. J., Further mass spectrometric analysis of organic polymers, Materials Quality Assurance Directorate, Ministry of Defence, Report No. 337, 1984.
86. Ardrey, R. E., Batchelor, T. M., and Smalldon, K. W., The analysis of acrylic fibres and white alkyd paints using pyrolysis probe mass spectrometry, Report No. 322, Home Office Central Research Establishment, 1979.
87. Wheals, B. B., Forensic applications of analytical pyrolysis, in *Analytical Pyrolysis,* Jones, C. E. R. and Camer, C. A., Eds., Elsevier, Amsterdam, 1977, 89.
88. Hughes, J. C., Wheals, B. B., and Whitehouse, M. J., Pyrolysis-mass spectrometry. A technique of forensic potential? *Forensic Sci.,* 10, 217, 1977.
89. Curry, C. J., Metropolitan Police Forensic Science Laboratory, unpublished results.
90. Hughes, J. C., Wheals, B. B., and Whitehouse, M. J., Pyrolysis mass spectrometry of textile fibres, *Analyst,* 103, 482, 1978.
91. Ardrey, R. E., Mullings, L. R., and Smalldon, K. W., Thermogravimetry-mass spectrometry — a new technique for the discrimination of closely related polymers, Report No. 274, Home Office Central Research Establishment, 1978.
92. Curry, C. J., Whitehouse, M. J., and Chalmers, J., Ultramicrosampling in infra red spectroscopy spectroscopy using small apertures, *Appl. Spectrosc.,* 39, 174, 1985.
93. Curry, C. J., Metropolitan Police Forensic Science Laboratory, unpublished work.

Chapter 7

MISCELLANEOUS FORENSIC APPLICATIONS OF MASS SPECTROMETRY

Jehuda Yinon

TABLE OF CONTENTS

I. INTRODUCTION

The use of mass spectrometry (MS) in forensic sciences is not limited to the applications described in the previous chapters. MS is a highly sensitive and specific analytical technique. It can therefore be used for the detection and identification of a wide variety of materials of forensic interest. A number of ionization techniques, such as electron impact (EI), chemical ionization (CI), field desorption, spark source ionization, laser ionization, and fast atom bombardment, make MS a method of choice for the analysis of both organic and inorganic samples.

The examples presented in this chapter are from those forensic applications published in the scientific literature. However, the number of possible uses of MS in the forensic laboratory is limited only by the imagination of the forensic mass spectrometrist.

II. MASS SPECTROMETRY OF TEAR GASES

Commercially available chemical protection sprays contain tear gases as active ingredients, together with a propellant and a suitable organic solvent. Lachrymatory compounds commonly used are chloroacetophenone (CN) *(1)*

o-Chlorobenzylidenemalononitrile (CS) *(2)*

CN and CS have been used as military tear gases.[1]

Another tear gas is capsaicin *(3)*, trans-8-methyl-*N*-vanillyl-6-nonenamide.

Capsaicin is extracted from the Cayenne pepper, *Capsicum*. The crude extract of the *Capsicum* fruit is oleoresin *Capsicum* and is a complex mixture of oils, waxes,

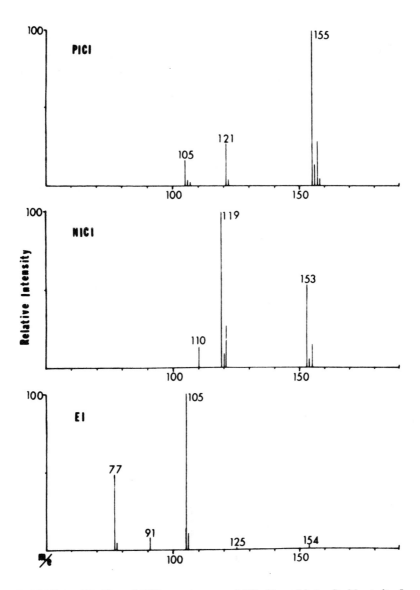

FIGURE 1. EI, CI, and NCI mass spectra of CN. (From Martz, R. M. et al., *J. Forensic Sci.*, 28, 200, 1983. With permission.)

colored materials, and several vanillyl amides called capsaicinoids.[2] Five capsaicinoids have been identified in naturally occurring *Capsicum:* capsaicin, dihydrocapsaicin (DC), nordihydrocapsaicin (NDC), homocapsaicin (HC), and homodihydrocapsaicin (HDC). Two of these, capsaicin and DC, constitute 80 to 95% of the total capsaicinoids in *Capsicum* extracts.

CN and CS are easily extracted from the liquid sprays with methanol. An aqueous sodium hydroxide extraction procedure is needed for sprays containing capsaicin.[1]

Articles of clothing can be analyzed for the presence of tear gas residue. CN can be identified by analyzing a heated vapor headspace sample of the cloth by GC/MS. A hexane wash of the cloth is needed for identification of CS and capsaicin.[1]

Mass spectra of these tear gases have been recorded by several groups.[1-6] The EI, CI, and negative-ion chemical ionization (NCI) mass spectra of CN are shown[6] in Figure 1. The base peak in the EI mass spectrum is the fragment ion at m/z 105,

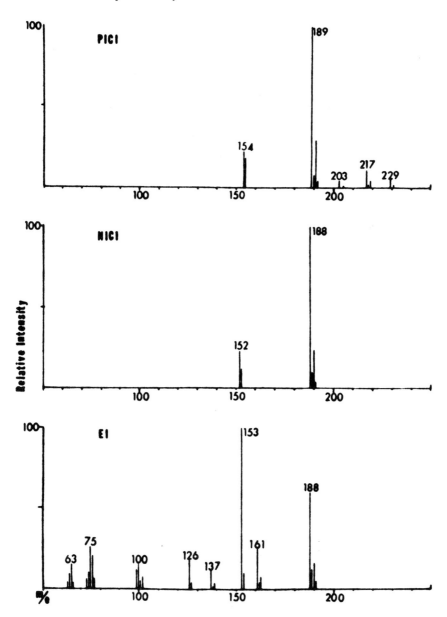

FIGURE 2.   EI, CI, and NCI mass spectra of CS. (From Martz, R. M. et al., *J. Forensic Sci.*, 28, 200, 1983. With permission.)

probably due to loss of (Cl + CH₂) from the molecular ion. The base peak in the methane CI-spectrum is the MH⁺ ion at m/z 155, which is accompanied by its isotopic ion at m/z 157. These two ions have a characteristic 3:1 chlorine 35/37 isotopic pattern. The NCI mass spectrum with methane as moderator gas has an abundant ion at m/z 153, probably due to (M − H)⁻. The base peak is at m/z 119. If we assume that this ion is due to (M − Cl)⁻, then the ion at m/z 121 is not an isotopic ion of Cl. The most sensitive method was found to be CI.

The EI, CI, and NCI mass spectra of CS are shown[6] in Figure 2. The base peak in the EI mass spectrum is at m/z 153 due to loss of Cl from the molecular ion.[5] An abundant molecular ion, together with its isotopic ion, are observed at m/z 188 and

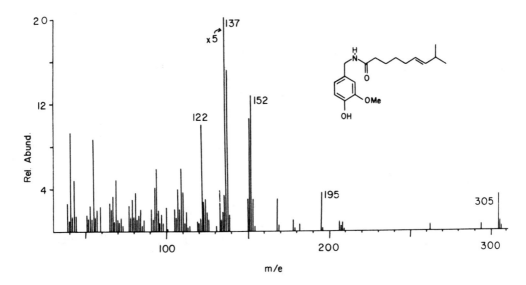

FIGURE 3. EI mass spectrum of capsaicin. (From Smith, R. M., *Int. Lab.*, 115, May/June, 1978. With permission.)

190, respectively. The base peak in the CI-methane spectrum is the MH⁺ ion at m/z 189 which is accompanied by its isotopic ion at m/z 191. The NCI mass spectrum with methane as moderator has its base peak at m/z 188, probably due to (M − H)⁻. The most sensitive method was found to be NCI.

The EI mass spectrum of capsaicin was obtained by GC/MS[1,4] (Figure 3). The mass spectrum is characterized by the base peak at m/z 137, due to the vanillyl ion, and by the molecular ion at m/z 305.

GC/MSEI spectra of TMS derivatives of capsaicin and its analogs DC, NDC, HC, and HDC were recorded at electron energies of 20 eV[3,7] and 70 eV.[2,3] A mass fragmentographic method[2,7] was used for the quantitative analysis of the TMS derivatives of the capsaicinoids, using the molecular ions for monitoring the various components.

III. FORENSIC DETECTION OF ART FORGERY BY MASS SPECTROMETRY

The value of an art object is primarily determined by its authenticity and not by its artistic quality. Many forgeries are made so skillfully that the art expert can easily be misled.[8] The application of scientific methods is therefore the only objective means of detecting these forgeries.

A. Authentication of Paintings

Authenticity judgements of paintings have been made by trace analysis of pigment elements and pigment impurities. Using spark source MS, detection of elements down to the parts per million range was possible.[9] The sample size necessary for analysis was in the milligram range.

The problem with this method is that the pigments may have been accidentally contaminated. Also, variations in the refining and smelting process could lead to significant differences between pigment batches.

A method which is unaffected by the conversion processes from raw material to pigments is the isotopic analysis of the main pigment elements like carbon, oxygen, sulfur, and lead. The isotopic composition of some of these elements may differ according to the geographic origin of the raw material, while others may provide chronological information. For example, the relative abundance of the sulfur isotopes ^{32}S/^{34}S in lapis lazuli from Afghanistan — the only ultramarine source known until the 19th century — differs by 0.52 from that of the ultramarine mined in the Chilean Andes.[8]

Another example[8] is the measurement of ^{18}O:^{16}O oxygen isotope ratios of native ochres, consisting largely of Fe_2O_3. Physical and chemical fractionation processes are responsible for the isotopic variations in oxygen. These ratios could indicate whether the pigment originates from the earlier Dutch sources or from the later Italian raw and burnt Sienna forms.

An additional application is the authentication of paintings by isotopic analysis of the white lead pigments.[8,10]

MS is obviously the ideal technique for the determination of small differences in stable isotope ratios.

High-resolution MS has been used for the analysis of media in Indian miniature paintings.[9] The medium, usually an exudate from a tree such as gum arabic, is first hydrolyzed into its basic components. The various gums are identified by GC and high-resolution MS.

B. Characterization of Ancient Objects by Lead Isotope Analysis

A large number of archeological objects contain lead in various amounts. Among these are glasses, glazes, cosmetics, leaded bronzes (e.g., bronze coins), silvers, and golds.[10,11] The isotopic composition of lead varies in nature because three of the four natural isotopes of lead are partly derived from the radioactive decay of ^{238}U, ^{235}U, and ^{232}Th into ^{206}Pb, ^{207}Pb, and ^{208}Pb, respectively. For example,[12] the ratio ^{208}Pb:^{206}Pb varies in nature over a range up to 5%. Such variations are the basis for lead isotope geochronology and geochemistry. As the isotopic composition of lead in ancient objects is unchanged through metallurgic or other processes, it may be used for the determination of the geological age or the geochemical origin of the object. By determining the isotopic composition of lead samples removed from ancient objects, and comparing them to the compositions of galena ores (lead sulfide) it becomes possible to determine the mining region from which the leads in those objects have come.[11]

Determination of isotope ratios of microgram quantities of lead has been done by thermal ionization MS.[13] The separation of lead from a wide variety of matrixes was done by anodic deposition and was obtained in the form of lead dioxide (PbO_2). The precision of the isotope ratio measurements was 0.1%.

Lead isotope ratio measurements have been used to determine the authenticity of ancient bronze coins.[8,11] For example,[8] for an ore sample collected from the famous ancient lead-mining district of Laurion, Greece (some 40 km from Athens) the following lead isotope ratios were obtained:

$$^{204}\text{Pb}: {}^{206}\text{Pb} = 0.0530$$

$$^{208}\text{Pb}: {}^{206}\text{Pb} = 2.0599$$

$$^{207}\text{Pb}: {}^{206}\text{Pb} = 0.8307$$

Figure 4 shows the lead isotope ratios of various galena ore sources and ancient objects.[8,11] Large differences in the isotopic ratios are observed.

Glass is another medium which can be characterized by isotopic analysis. Two important types of ancient glasses, red and yellow opaques, often contain appreciable quantities of lead as an additive.[8,10] The red opaques owe their color to the presence of

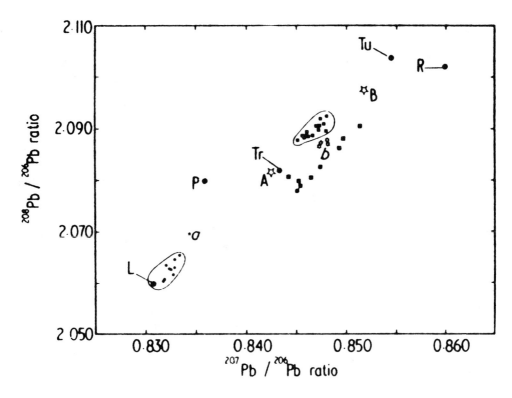

FIGURE 4. Lead isotope ratios of various galena ore sources and ancient objects. Galena ore sources: L, Laurion, Greece; P, Populonia (now Campiglia Marittima, Italy); Tr, Trabzon, Black Sea Coast; Tu, Bottino, Tuscany; R, Rio Tinto, Spain. Bronze coinage: Greek (330 to 32 B.C.), circles; Roman A.D. 111 to 308, small squares. (a) Imperial coin of Athens, 27 B.C. to A.D. 14, with a bust of Athena on the obverse. (b) (Small "starred" group). Four provincial Roman coins minted at London, c. A.D. 307 to 296. Lead pigs from Roman-British sites: large squares. (A) Isotope data from Julia Domna bronze portrait bust. (B) Coin of Maxentius struck in Rome c. A.D. 307 but a suspected contemporaneous forgery. (From Fleming, S. J., *Authenticity in Art: The Scientific Detection of Forgery,* The Institute of Physics, London, 1975. With permission.)

cuprite crystallites precipitated in the glass under reducing heating conditions. This chemical reaction is enhanced by the addition of more than 5% lead oxide. The yellow color is obtained more directly by addition of one of two opacifiers, lead diantimonate ($Pb_2Sb_2O_7$) or lead stannate ($PbSnO_3$). For example, ancient Egyptian glass has been characterized[8] by its exceptionally low isotope ratios:

$$^{208}Pb: {}^{206}Pb \cong 2.025$$

$$^{207}Pb: {}^{206}Pb \cong 0.811$$

Direct analysis of lead-containing glasses and glazes was carried out by fast atom bombardment (FAB) MS without a glycerol matrix.[14] Lead isotope ratios were measured to a relative precision ranging from 0.3 to 1%. This precision was good enough to determine broad distinctions between samples but not good enough to pinpoint the exact location where the lead was mined.

C. Characterization of Ancient Glass by Oxygen Isotope Analysis

Oxygen isotope analysis provides a separate method for ancient glass analysis because it refers to the principal silica-bearing ingredient rather than to an additive.[8,10]

Oxygen is actually the most abundant element in glasses, accounting for some 45 to 50% by mass of the glass.

By convention, the ^{18}O content is usually expressed in terms of δ, which is the deviation in parts per thousand of the ^{18}O content of the sample from that of an accepted standard designated as standard mean ocean water (SMOW):

$$[^0/_{00}]\ \delta = \left[\frac{(^{18}O:\ ^{16}O)\ \text{sample}}{(^{18}O:\ ^{16}O)\ \text{ocean}} - 1 \right] \times 1000$$

Positive values of δ indicate an excess of ^{18}O over the standard. As an example, the typical δ values of some ancient glasses are as follows:[8,10]

18th Dynasty glasses from Egypt (c. 1350 B.C.): $\delta = 16.1^0/_{00}$
Glasses from Nimrud, Mesopotamia (7th century B.C.): $\delta = 22.0^0/_{00}$
Glasses from Jelemie, West Galilee (4th century A.D.): $\delta = 14.1^0/_{00}$

Variations in $^{18}O:^{16}O$ and $^{13}C:^{12}C$ isotope ratios in Greek marbles have been used as a method for the unique characterization by locality.[15] ^{18}O varies by at least 2% in marbles from different areas. The analytical precision of the mass spectrometric isotope ratio measurement[15] was $\pm 0.05^0/_{00}$.

D. Characterization of Ancient Papyri

Ancient Egyptian papyri were examined by pyrolysis/MS in order to determine the age, composition, and processing of this ancient paper.[16]

The main components of the papyrus plant are cellulose, hemicellulose, and lignin. It is therefore of interest to compare the mass spectrum of the papyrus with the mass spectra of its components.

The amount of sample necessary for analysis was 10 to 50 μg. The sample was heated to 500°C in the pyrolyzer which was connected directly to the mass spectrometer. An integrated mass spectrum of all the volatile pyrolyzates was recorded during the complete pyrolysis procedure. Mass spectra recorded during the pyrolysis of cotton cellulose, lignin, and the ancient papyrus Westcar are shown in Figure 5. The mass spectrum of cellulose indicates the presence of levoglucosane (1,6-anhydro-β-D-glucopyranose) (M^+ at m/z 162) and its decomposition products at m/z 144 and m/z 126. The spectrum of lignin shows the presence of coniferyl alcohol, a lignin monomer, (m/z 180), as well as other substituted lignin phenols.

In order to investigate the effect of age on cellulose structure, mass spectra of pyrolysis products of various ages were recorded. The spectra showed a decrease or disappearance of the peaks at m/z 162 (M^+) and m/z 144 $(M - H_2O)^+$ with increasing age, indicating that levoglucosane, the monomeric building block of cellulose, undergoes single or multiple dehydration with time and is eventually transformed to levoglucosenone (m/z 126). The result of this dehydration can be observed in the mass spectrum of the papyrus Westcar (Figure 5).

IV. FORENSIC APPLICATIONS OF SPARK SOURCE MASS SPECTROMETRY

A. Differentiation of Bullets

Matching of evidence bullets with a suspect's gun can be accomplished by comparing the rifling marks on the evidence bullets with those on test bullets fired from the sus-

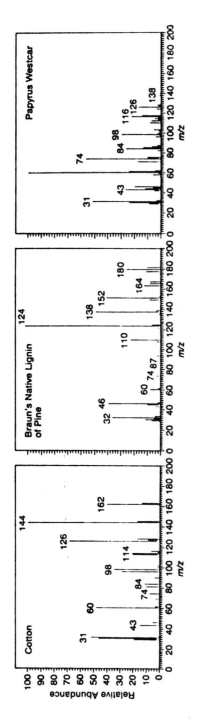

FIGURE 5. Pyrolysis/mass spectra of cotton cellulose, lignin, and the ancient papyrus Westcar. (From Wiedemann, H. G. and Bayer, G., *Anal. Chem.*, 55, 122A, 1983. With permission.)

Table 1

ANALYSES OF SOME WINDOW GLASSES

| Sample | Refractive index | Density | Concentrations in ppm | | | | | | |
			Potassium	Iron	Strontium	Zirconium	Barium	Cerium	Lead
8	1.5184	2.5046	1270	920	77	150	52	4.5	4.5
7	1.5184	2.5040	1100	1060	84	175	53	4.0	5.0
13	1.5185	2.5040	219	680	38	57	12.4	1.5	4.1
6	1.5184	2.5058	180	700	128	87	20	2.6	3.7
11	1.5186	2.5057	174	800	65	74	15.2	3.0	—
10	1.5186	2.5055	147	890	49	61	13.6	2.6	—
9	1.5186	2.5052	139	740	57	73	12.7	2.7	—
5	1.5183	2.5046	118	920	39	162	4.5	2.6	3.3

From Haney, M. A., *J. Forensic Sci.*, 22, 534, 1977. With permission.

pect's gun. This method cannot be used with the rifling marks on the bullet found at the scene of the crime are obliterated, or when the gun is not available for test firing. If the elemental composition of the evidence bullets is found to be identical to that of bullets found in the possession of the suspect, this may be used as circumstantial evidence against him.

Spark source MS was used[17,18] to characterize bullets according to their elemental composition. About 15 to 20 mg of bullet lead is required for a complete elemental analysis at the 0.1-ppm level range. Three techniques were used: photographic detection, which is the most versatile, revealing which elements are quantitatively useful, electrical detection by magnetic peak switching, which provides better speed and precision for a limited set of elements, and isotope dilution, which enables absolute determination of elemental composition without the need of standards.

Among a list of 26 elements commonly found in bullets at the 0.1-ppm level and above, only the following ones were found[18] to be suitable for quantitative characterization: S, Cu, As, Se, Ag, Cd, Sn, Sb, Te, Hg, Tl, and Bi.

Approximately five bullets from each of ten boxes of different types of bullets were analyzed.[18] It was found that the bullets in each box do not have uniform elemental compositions (with the exception of antimony [Sb]). However, there were groups of bullets within each box which had similar compositions. Although the accuracy was about 30%, it was found that spark source MS was superior to the more precise method of neutron activation analysis because of the larger number of elements which could be determined.

B. Differentiation of Window Glasses

Isotope dilution spark source MS was used[19] for the characterization of window glasses. Although isotope dilution is unnecessary for direct comparison of evidence, absolute measurements enable the accumulation of a database as a reference for comparison.

Large glass fragments were ground into powder, and 10 mg of the sample was dissolved in sulfuric and hydrofluoric acid. The dissolved samples were spiked with 50 to 500 $\mu\ell$ of separated isotope solutions. The concentration and volume of added isotope spikes were adjusted to give an isotopic ratio of approximately unity in the final solution. The added isotopes included ^{41}K, ^{57}Fe, ^{87}Rb, ^{86}Sr, ^{91}Zr, ^{137}Ba, ^{142}Ce, and ^{207}Pb.

Table 1 shows the results of a series of window glass samples.[19] The relative standard deviation was 6%.

The elements potassium, rubidium, strontium, zirconium, and barium were found to be homogeneously distributed within window panes and therefore could be useful for the characterization of window glasses. Iron, lead, and cerium were of little use because of their inhomogeneity within the panes.

C. Trace Element Analysis in Hair and Fingernails

Spark source MS has been used to study human hair and fingernail samples with respect to their total inorganic content.[20] Elimination of the organic content of the samples was necessary in order to produce a simple and easily interpretable spectrum. This was done by an ashing procedure.

Samples of hair studied showed that 25 to 30 elements could be determined for comparison, allowing an assignment of a relative degree of similarity.

The storage site characteristics of human hair are of value in monitoring toxicological intake of certain elements. For example, the ingestion of mercury, arsenic, or lead may produce abnormally high concentrations in hair. A case sample of suspected metal poisoning revealed a very high titanium level and was traced back to a daily occupational exposure in an industrial plant.

Human fingernails act in a similar manner to hair in terms of trace element storage capability. Table 2 shows a typical trace element analysis of two human fingernail samples.[20]

A case of arsenic poisoning showed a fivefold increase of arsenic in both nails and hair with regard to a normal level.

V. FORENSIC MASS SPECTROMETRY OF VARIOUS COMPOUNDS

A. Analysis of Dye Residues

Exploding money packets are a tool to help the investigation of robberies. These devices are remotely activated to deflagrate or explode after the robber leaves the bank, rupturing the container in which the money is being held and ejecting an aerosol of red dye and tear gas.[6] The red dye will stain the robber and the money, while the tear gas is supposed to cause the abandonment of the money. The identification of the red dye on clothing, on automobile upholstery, or on money will sometimes link a suspect to a bank robbery.

GC/MS has been used to detect and identify[6] submicrogram quantities of a commonly used red dye, 1-methylaminoanthraquinone (MAAQ) *(4)*.

Extraction was done with chloroform or acetone; 1 μl of the solution was injected into the GC/MS. Figure 6 shows the EI, CI, and NCI mass spectra of MAAQ. The EI spectrum has a molecular ion base peak at m/z 237 and a large number of fragment ions. The CI-methane spectrum consists mainly of the MH^+ ion at m/z 238 while the NCI spectrum, with methane as moderator, consists of the $(M - H)^-$ ion at m/z 237. The most sensitive method was found to be NCI.

Table 2

ANALYSIS OF TWO
FINGERNAIL SAMPLES
FROM DIFFERENT
SUBJECTS

	Weight (μg/g)	
Element	Sample 1	Sample 2
Pb	7.2	11.0
Ce	0.34	—
La	0.25	—
Ba	2.8	2.0
Sn	2.1	4.3
Cd	1.0	2.6
Ag	—	1.0
Mo	2.7	—
Zr	0.44	—
Sr	—	0.60
Rb	—	1.9
Se	1.7	0.83
As	0.71	0.48
Ge	1.0	1.6
Zn	124	187
Cu	25	59
Fe	92	74
Mn	2.2	1.2
Cr	2.8	3.4
Ca	690	1600
K	2900	5400
S	4700	4700
P	200	392
Mg	59	170
Na	850	578

From Harrison, W. W. et al., *J. Assoc. Off. Anal. Chem.*, 54, 929, 1971. With permission.

B. Mass Spectrometry of Pencil Marks

Since its introduction during the 16th century, the pencil has become the most widely used writing instrument throughout the world.[21] Today, the number of documents that bear some form of pencil writing is remarkable. Some of these penciled writings later come to possess evidentiary value. The identification of a particular type of pencil lead used in the execution of a specific document can be of great importance.

The term "lead" pencil is a misnomer in that the pencil contains no lead but rather is composed of three primary ingredients in varying proportions: graphites, clays, and waxes.[21]

EIMS of pencil marks was carried out in order to study the possibility of discrimination between various pencil cores.[22] The samples were introduced into the mass spectrometer by a direct-insertion probe fitted with a glass sample tube. Pencil marks were removed from paper by scraping the surface with a blade to produce a ball of fibers that was then placed inside the sample tube.

Mass spectra of pencil cores are essentially mixed spectra of the waxes used in the manufacturing process of the pencil lead. A certain number of characteristic ions (m/z 219, 227, 229, 239, 241, 256, 257, and 264) were monitored by multiple ion detec-

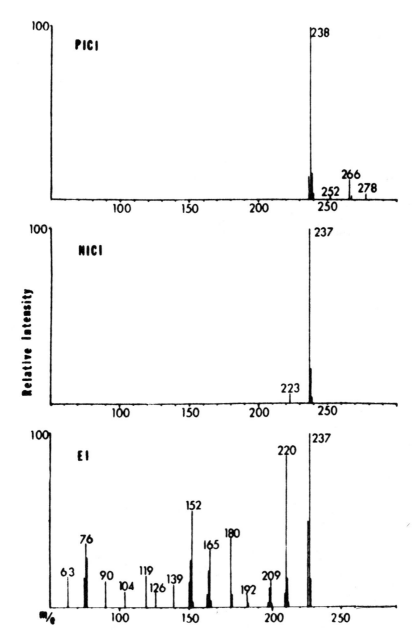

FIGURE 6.   EI, CI, and NCI mass spectra of MAAQ. (From Martz, R. M. et al., *J. Forensic Sci.*, 28, 200, 1983. With permisison.)

tion.[22] It was found that this group of ions enabled the possibility of discrimination between various pencil cores and suffered less from background interference than lower mass ions. Mass spectra of samples were compared with mass spectra of unmarked paper samples in order to eliminate the contribution of traces of volatile organics in the paper.

A small collection of 17 pencils of various brand names and hardness designations was examined.[22] The marks made by these pencils could be divided, according to the mass spectral ions, into four groups. The pencil marks remained easily distinguishable by MS for several months. However, in attempting to apply the method to some forensic documents, several problems became apparent. When the paper on which a mark was made was soiled or had been much handled, the mass spectrum of the mark was obscured by skin oils or other materials on the fibers scraped up with the mark.

REFERENCES

1. Nowicki, J., Analysis of chemical protection sprays by gas chromatography/mass spectrometry, *J. Forensic Sci.*, 27, 704, 1982.
2. Fung, T., Jeffery, W., and Beveridge, A. D., The identification of capsaicinoids in tear-gas spray, *J. Forensic Sci.*, 27, 812, 1982.
3. Lee, K.-R., Suzuki, T., Lobashi, M., Hasegawa, K., and Iwai, K., Quantitative microanalysis of capsaicin, dihydrocapsaicin and nordihydrocapsaicin using mass fragmentography, *J. Chromatogr.*, 123, 119, 1976.
4. Smith, R. M., Some applications of GC/MS in the forensic laboratory, *Int. Lab.*, 115, May/June, 1978.
5. Avdovich, H. W., By, A., Ethier, J.-C., and Neville, G. A., Spectral identification of a lachrymatory exhibit as CS, *Can. Soc. Forensic Sci. J.*, 14, 172, 1981.
6. Martz, R. M., Reutter, D. J., and Lasswell, L. D., III, A comparison of ionization techniques for gas chromatography/mass spectroscopy analysis of dye and lachrimator residues from exploding bank security devices, *J. Forensic Sci.*, 28, 200, 1983.
7. Iwai, K., Suzuki, T., and Fujiwake, H., Simultaneous microdetermination of capsaicin and its four analogues by using high-performance liquid chromatography and gas chromatography-mass spectrometry, *J. Chromatogr.*, 172, 301, 1979.
8. Fleming, S. J., *Authenticity in Art: The Scientific Detection of Forgery*, The Institute of Physics, London, 1975.
9. Johnson, B. B. and Cairns, T., Art conservation: culture and analysis, *Anal. Chem.*, 44, 24A, 1972.
10. Brill, R. H., Lead and oxygen isotopes in ancient objects, *Phil. Trans. R. Soc. London A*, 269, 143, 1970.
11. Brill, R. H. and Shields, W. R., Lead isotopes in ancient coins, *Roy. Numismatic Soc. Spec. Publ.*, 8, 279, 1972.
12. Gale, N. H. and Stos-Gale, Z. A., Bronze Age copper sources in the Mediterranean: a new approach, *Science*, 216, 11, 1982.
13. Barnes, I. L., Murphy, T. J., Gramlich, J. W., and Shields, W. R., Lead separation by anodic deposition and isotope ratio mass spectrometry of microgram and smaller samples, *Anal. Chem.*, 45, 1881, 1973.
14. Dolnikowski, G. G., Watson, J. T., and Allison, J., Direct determination of metals in archaeological artifacts by Fast Atom Bombardment Mass Spectrometry, *Anal. Chem.*, 56, 197, 1984.
15. Craig, H. and Craig, V., Greek marbles: determination of provenance by isotopic analysis, *Science*, 176, 401, 1972.
16. Wiedemann, H. G. and Bayer, G., Papyrus, the paper of ancient Egypt, *Anal. Chem.*, 55, 122A, 1983.
17. Haney, M. A. and Gallagher, J. F., Elemental analysis of bullet lead by spark source mass spectrometry, *Anal. Chem.*, 47, 62, 1975.
18. Haney, M. A. and Gallagher, J. F., Differentiation of bullets by spark source mass spectrometry, *J. Forensic Sci.*, 20, 484, 1975.
19. Haney, M. A., Comparison of window glasses by isotope dilution spark source mass spectrometry, *J. Forensic Sci.*, 22, 534, 1977.
20. Harrison, W. W., Clemena, G. G., and Magee, C. W., Forensic applications of spark source mass spectrometry, *J. Assoc. Off. Anal. Chem.*, 54, 929, 1971.
21. Cain, S., Cantu, A. A., Brunnelle, R., and Lyter, A., A scientific study of pencil lead components, *J. Forensic Sci.*, 23, 643, 1978.
22. Zoro, J. A. and Totty, R. N., The application of mass spectrometry to the study of pencil marks, *J. Forensic Sci.*, 25, 675, 1980.

INDEX

G

H

I